Cryogenic Processes and Equipment 1982

T. H. K. Frederking, M. J. Hiza
W. M. Toscano and L. A. Wenzel, editors

R. J. Adler
A. S. Adorjan
Kenneth Anselmo
V. D. Arp
J. T. Bernstein
M. J. P. Bogart
C. B. Brosilow
W. R. Brown
Shlomo Caspi
Nirmal Chatterjee
E. H. Christensen
D. B. Crawford
D. E. Daney
Victor Dilline
Ken Dunn
F. J. Edeskuty
J. F. Ely
G. W. Fenner
J. F. Flower
T. H. K. Frederking
N. C. Gardner
P. J. Giarratano
William Gilbert
Geoffrey Green
S. E. Handman
G. G. Haselden
William Hassenzahl
J. W. Hempseed
D. J. Hersh
R. E. Hise
M. J. Hiza
J. M. Kaufman
Ifiyenia Kececioglu
A. J. Kidnay
Y. I. Kim
P. A. Lessard
L. G. Massey
A. J. Munday
N. A. Olien
P. S. O'Neill
George Patton
F. W. Pirtle
A. Powell
Jeb Rechen
R. C. Reid
A. Stephan
W. G. Steward
Helmut Springmann
D. W. Studer
Clyde Taylor
Richard Warren
K. D. Williamson
K. J. Yang

AIChE Symposium Series

Number 224 | 1983 | Volume 79

Published by

American Institute of Chemical Engineers

345 East 47 Street | New York, New York 10017

AIChE shall not be responsible for statements or opinions advanced in papers or printed in its publications.

Library of Congress Cataloging in Publication Data

Main entry under title:
Cryogenic processes and equipment, 1982.

(AIChE symposium series; no. 224, 1983, v. 79)
1. Low temperature engineering—Congresses. I. Frederking, T. H. K. II. American Institute of Chemical Engineers. III. Series: AIChE symposium series; no. 224.
TP480.C68 1983 621.5'9 83-11314
ISBM 0-8169-0249-6

Printed in the United States of America by
Lew A. Cummings Co., Inc.

FOREWORD

Joint Cryogenics Symposia organized by Committees of AIChE, ASME and IIR have taken place since 1976 in those years when the Cryogenic Engineering Conference did not meet (i.e. in even years). This joint effort has been quite successful replacing small group gatherings by an efficient exchange of information in areas of applied cryogenics.

The following symposia have been held so far:

1976 New York, host society ASME
1978 Miami, host society AIChE
1980 San Francisco, host society ASME
1982 Los Angeles, host society AIChE

The present volume constitutes a collection of papers presented in Los Angeles in 1982 in areas of general cryogenic processes and equipment including support systems at relatively high temperatures. Further, progress in the Helium-4 technology all the way down to 2 K is reported. In addition, advances in understanding of transport phenomena and equilibration processes are documented.

The editors wish to acknowledge the efforts of all who contributed to this volume: all the authors and symposia speakers, the session chairmen, participants, discussers and reviewers. Finally, members and officers of the cryogenic committees of the three participating societies. Last, but not least, the National organization of the AIChE, the Publications Director Larry Resen, and his editorial staff, in particular Maura Mullen.

T.H.K. Frederking
M.J. Hiza
W.M. Toscano
L.A. Wenzel

CONTENTS

THERMOPHYSICAL PROPERTIES AND TRANSPORT PHENOMENA

RECENT DEVELOPMENTS TOWARD A H_2 ECONOMY

This paper highlights the current areas of hydrogen research and application interest on a worldwide basis. Developments discussed are mainly in the areas of hydrogen production, safety, and in fuel applications. The latter is limited to automotive and aircraft applications.

K. D. WILLIAMSON, JR.
and
F. J. EDESKUTY

Los Alamos National Laboratory
Analysis and Assessment Division
Energy Technologies Group

RESEARCH AND DEVELOPMENT: THERMOCHEMICAL AND SAFETY

In the area of research and development, two topics are of prime interest--thermochemical production methods and safety.

The thermochemical work is being conducted in the United States at the Los Alamos National Laboratory, the Lawrence Livermore National Laboratory, the Institute of Gas Technology, General Atomic Co., and the Westinghouse Electric Corp.; in Italy at Ispra; in Germany at the Aachen Institute; and in Japan. While thermochemical processes possess theoretically high overall efficiencies (70 to 75% versus 40% for electrolysis) and therefore hold the promise of lower hydrogen cost, several generic problems exist. These include efficient heat transfer and heat recuperation, loss of intermediate materials, and the separation of chemical species.

The most advanced thermochemical cycles are the hybrid H_2SO_4 cycle, the H_2SO_4/HI cycle, and the H_2SO_4/HBr cycle. The latter cycle is being developed at Ispra. They have a laboratory scale unit that has produced 100 l/h of hydrogen.

Unfortunately, these cycles appear to have attainable overall efficiencies on the order of 40%. It is the authors feeling that because of the complexities of thermochemical processes such processes must have clear-cut advantages over competing technologies (e.g., electrolysis) if they are to have an impact in the marketplace. At this point, it is not clear whether or not these technologies will have such advantages (1).

In the area of safety, the recent efforts have been concerned with large-scale spill tests (2) and with hydrogen generation in light-water reactor loss-of-coolant accidents. The spill tests were conducted under the sponsorship of NASA at White Sands, New Mexico, and simulated ruptures of large storage vessels. Up to 1500 gal of LH_2 were spilled in 35 s. Cloud contours, composition, and history were determined. In rapid spills, violent turbulence was generated which resulted in fast mixing with air to concentrations below the lower flammability limit. The cloud quickly became bouyant with a rise rate of 0.5 to 1.0 m/s. In the prolonged spill cases, the ground at the spill site cooled significantly, greatly reducing the heat source for inducing violent vaporization. As a result, prolonged ground level, flammable cloud travel could occur and in fact was observed.

The LWR studies are, of course, a result of the Three Mile Island event. The major United States-sponsor of the work is Electric Power Research Institute with support from Canada, Sweden, and Taiwan (3).

The purpose of the program is to study H_2 combustion and control in various environments that could occur following a postulated, severe nuclear accident. The major concerns are containment vessel integrity in the presence of pressure generated by a H_2 burn, equipment survivability due to temperature transients, and containment and equipment survivability due to local H_2 detonations, if any. The mitigation strategy of choice is controlled combustion because it is simple, passive, and apparently reliable. Many studies are underway or complete in the areas of combustion, ignition effectiveness, mixing, and model development.

In small scale tests (vessels 0.02 to 850 m^3), the major findings to date are: (1) deliberate ignition devices are effective in various environments; (2) during burns, maximum pressure increases of less than two atm have been observed; (3) no detonations have been observed; (4) equipment has survived multiple combustion events; and (5) computer codes and models for combustion and mixing require further improvements.

A series of large-scale demonstration tests is scheduled to begin late in the spring of 1983 at the site of the Nuclear Rocket Development Station in Nevada. These tests will utilize a 2100 m^3 LH_2 dewar (approximately one tenth actual LWR containment vessel size) to verify the small-scale test data, to examine equipment survivability, and to test deliberate H_2 ignition during simulated degraded-core accidents. The dewar system is being modified to provide up to 240 gpm water spray, up to 17 000 lb/h of 150 psi steam, 1 to 7 lb/min of hydrogen, and up to 600 Ft^3/min of compressed air. Both static and dynamic tests are planned.

APPLICATION ORIENTED PROGRAMS: VEHICLES AND AIRCRAFT

The work on automotive applications is centered in the United States, Canada, Japan, and Germany. Among the major efforts in the United States, the work at the Los Alamos National Laboratory, the University of Miami, and Solar Turbine International will now be summarized.

Routine driving and refueling tests with liquid hydrogen using a 1979 Buick were recently completed at Los Alamos (4). The program, conducted jointly with the DFVLR of Germany, accomplished 65 refueling operations. A total of 2300 miles were driven with H_2 as the fuel. No problems of consequence were encountered. The LH_2 dewar fit into the trunk space, had a capacity of 155 liters and a boil-off rate of 4%/day, which permitted a 60-hour lockup before reaching the maximum allowable pressure of 65 psia. The work resulted in an expanded data base for vehicle operation, indicated that safe refueling can be carried out by simi-skilled workers, demonstrated that the dewar system needs further development to reduce the heat leak, and found that a continuing operational safety program would be needed in parallel with the technology development.

At the University of Miami, work on H_2 engine designs and data base documentation are continuing. A recently completed portion of the program covered 19 different engine configurations using basic stock Toyota engine components. Following the lead of the Japanese Musashi group, they confirmed that NO_x emissions are greatly reduced (by factors of up to 20) with post intake valve closure injection and glow plug ignition (diesel mode of operation).

The effort at Solar Turbine International has been on the development of lightweight Mg hydrides. Their goal of 200°C dissociation temperatures has not as yet been achieved. However, they have developed hydrides with H_2 mass fractions of 3 to 4.5%.

On the international scene, the Urban Transportation Development Corp. in Canada recently terminated a \$6 x 10^6 H_2-fueled transit bus research and development effort. Great strides have been made in Japan at the Musashi Institute in the development of low-emission diesel-cycle engines. In Germany, the DFVLR has a dual fuel program (LH_2 and gasoline) with BMW and have taken the lead in developing LH_2 refueling stations (used in the Los Alamos program). In addition, Daimler-Benz is operating a fleet of hydride-storage hydrogen-fueled buses, sedans, and station wagons in Berlin. The source of the hydrogen is town gas (5).

International interest in hydrogen-fueled aircraft continues through the International Hydrogen in Air Transportation Ad Hoc Executive Group (6). Countries represented include the United States, Canada, France, Germany, and Japan. From the United States perspective, synjet from oil shale appears to be the best long-term option. However, countries with no oil resources take a more positive veiw of hydrogen. Since we are a major supplier of international aircraft, our aircraft industry (Lockheed-California Co. in particular) maintains an active interest in this application (6).

UTILIZATION PROSPECTIVE

It is the authors opinion that large-scale usage of hydrogen for transportation in the United States is many decades away. It is difficult for a new fuel to enter the market when no or little infrastructure exists and when one is competing with conventional fuels from oil or essentially identical synthetic fuels produced from oil shale and coal. Because of the many advantages hydrogen offers the aircraft industry and because of the limited infrastructure required, this application will probably be the first major transportation application in the United States. In the province of Ontario, Canada, the Institute of Hydrogen and Eletrochemical Systems has budgeted $\$10 \times 10^6$ to investigate the production and use of hydrogen in the province. The hydrogen would be produced by electrolysis using electricity from their Candu reactors. The motivation behind their program is the desire to become independent of western province-supplied oil. Finally, hydrogen might well enter the transportation scene in a major way in a country such as Costa Rica. They have abundant low-cost hydoelectricity available for hydrogen prodution, essentially no indigenous oil supply, and would require a minimum infrastructure to utilize hydrogen as a fuel.

LITERATURE CITED*

1. F. J. Edeskuty, "Theromochemical Production of Hydrogen," Los Alamos National Laboratory report LA-UR-82-1227 Los Alamos, NM (1982).

2. R. D. Witcofski and J. G. Chirivella, "Experimental and Analytical Analyses of the Mechanisms Governing the Dispersion of Flammable Clouds Formed by Liquid Hydrogen Spills," Hydrogen Energy Progress-IV, 4, p. 1659, T. N. Veziroglu, W. D. Van Vorst, J. H. Kelley, eds., Proceedings of the World Hydrogen Energy Conference IV, Pasadena, CA, June 13-17, 1982.

3. L. Thompson, "EPRI Research on Hydrogen Combustion and Control for Nuclear Reactor Safety," ibid, 4, p. 1675.

4. W. F. Stewart, "Operating Experience with a Liquid-Hydrogen Fueled Buick and Refueling System," ibid, III, p. 1071.

5. W. J. D. Escher, "Hydrogen as an Automotive Fuel: Worldwide Update," presented at the 3rd Non-Petroleum Vehicular Fuels Symposium, Arlington, VA, October 12-14, 1982.

6. G. D. Brewer, "Fuel for Future Transport Aircraft," ASME paper 81-HT-80, presented at the 20th Joint ASME/AIChE National Heat Transfer Conference, Milwankee, WI, August, 1981.

*Some of these references might be difficult to obtain. Further assistance may be obtained from the authors.

MEDIUM PURITY OXYGEN PRODUCTION AND REDUCED ENERGY CONSUMPTION IN LOW TEMPERATURE DISTILLATION OF AIR

J. R. FLOWER
G. G. HASELDEN
University of Leeds,
Leeds LS2 9JT., UK
J. W. HEMPSEED
Air Products, Ltd.,
Walton-on-Thames, UK
A. C. POWELL
R. M. Parsons & Co. Ltd.,
Brentford, UK

Air separation by cryogenic distillation to produce a medium purity (95%) gaseous oxygen product has been investigated by computer simulation with supporting experiments. The use of several condenser-reboilers to distribute the thermal flux between two distillation columns allows significant reduction in the compressor power consumption and capital cost. Several versions of a basic flowsheet have been investigated by a detailed rating model simulation of the columns and associated heat exchangers. Data for stage pressure drops and efficiencies have been assumed from current industrial practice. Data specifying the rate of heat transfer between the condensing vapour and evaporating two phase froths have been estimated from previous work and confirmed in preliminary laboratory experiments. The effects of uncertainties in data have been assessed. The results support the view that multiple thermal links offer scope for significant process improvement.

J.W. Hempseed and A.C. Powell contributed to this work during postgraduate studies at the University of Leeds.

Introduction

Distillation of air to produce pure oxygen and nitrogen is a well established process which has been optimised to a very high degree. In the past, most attention has been paid to the design of processes for pure oxygen and nitrogen (99%). In the future demand for medium purity oxygen or oxygen enriched air will provide openings for different thermodynamic cycles which offer scope for energy savings. Since the main energy user, the air compressor, is smaller, its contribution to the capital cost of the process will also decrease. This must be balanced against the need for novel designs of distillation tray and more complex piping connections between the distillation columns.

This paper describes simulation investigations of a new cycle and outlines experiments to check assumed data in the calculations. The principle lying behind the new cycle is the use of distributed reflux and reboil in the distillation columns to reduce losses due to thermodynamic irreversibility. In the past the concept was outlined in terms of conceptual designs. The present work is a development using improved equipment designs. Some cases considered use of existing items while others illustrate the further savings to be made using new distillation trays with integral heat transfer surfaces.

The Existing Process

Separation of air into oxygen and nitrogen vapour products by the existing low temperature distillation process uses a compressor for the air feed, a main heat exchanger in which the superheat of the air feed, about 200 K, is removed by cooling against the product vapour streams, and two distillation columns with associated interchangers. In the first or high pressure (HP) column shown in Figure 1, part of the air vapour feed is separated into a rich oxygen liquid and a nitrogen rich overhead product by rectification. The heat liberated on condensation of the nitrogen rich vapours at the top of the HP column is transferred to an evaporating oxygen rich liquid produced at the base of the second or low pressure (LP) column. The evaporation creates a vapour oxygen product which returns to ambient conditions via the main heat exchanger and a reboil vapour stream which enables the distillation processes in the LP column to occur. The overhead liquid product from the HP column is used as reflux to the top stage of the LP column allowing a nitrogen rich vapour product to return to ambient conditions via the main air feed cooler.

As the difference in normal boiling points of oxygen and nitrogen is about 18 K, the main condenser-reboiler linking the two columns operates at a vapour-side pressure of 5.5 to 6.0 bars and a liquid-side pressure of 1.5 to 1.6 bars. The liquid-side pressure depends on the pressure drops across the LP column, the main air cooler and the required delivery pressure to the consuming process. The air feed compressor delivery pressure is set by the vapour-side pressure increased by the pressure losses across the HP column and the main air cooler. Since the raw material costs are essentially zero, the cost of the products arises from the compression work and the capital charges. Improvement of the existing process therefore places great stress on minimising irreversible losses in the heat exchange and separation equipment to reduce the energy consumption of the compressor and hence its capital cost.

The main factor determining the compressor load is the requirement that a maximum purity nitrogen vapour at the top of the HP column be condensed by heat transfer to an evaporating oxygen liquid product at the base of the LP column. This choice of compositions means that the temperature difference across the exchange surface will be a minimum at a given set of pressures, or that the air feed pressure will be a maximum for a given feasible temperature difference in the reboiler-condenser.

The Proposed process

The main change proposed is to distribute the condensing and evaporating loads over a number of smaller condenser-reboilers or thermal links, (5)(6) as shown in Figure 2. At each link, the difference in composition between the condensing and evaporating streams is smaller and so the operating pressure of the HP column may be reduced for feasible temperature differences across the heat transfer surfaces. Separation requirements in both columns place complex constraints on the allowed distribution of the total heat load. This then requires detailed rating calculations to be made on the entire flowsheet.

As a simple illustration, it may be noted that the difference in temperature between the top and bottom of the HP column is usually about 4 K due to the composition differences. Consequently if part of the total evaporation load were met by condensation of a vapour sidestream from the base of the HP column, the temperature difference across that surface would increase by about 4 K, or the operating pressure in the HP column could be reduced by about 1 bar. Other factors serve to change the actual values but the principle illustrated remains valid.

Simulation strategy

The separation process consists of three parts, the two columns and the associated heat exchangers by which the cold is recycled within the distillation step. The compressor and main air feed cooler, e.g. reversing heat exchanger or regenerator, do not interact significantly with the separation step and so they are simulated separately. A rating approach was adopted in setting up mathematical models of the process. As a simulation project progresses, it is to be expected that the models will become more detailed and that optimization will proceed by evolution through a series of case studies. The three parts of the separation process were modelled by three main programs whose input and output were organised and modified interactively by a fourth, executive, program.

The high pressure column calculation included the external condensers supplied by the vapour sidestreams. Once a set of heat loads had been specified, a performance calculation gave output data for the two product streams which passed through various heat exchangers before entering the low pressure column. The low pressure and heat exchanger programs were closely coupled and these models were solved iteratively to give consistent overall heat balances as well as detailed profiles within the low pressure column. In particular, the oxygen product evaporator in which heat is removed from the condensing air feed to the low pressure column is responsible for strong interactions between the heat exchanger and LP column programs.

Once the LP and heat exchanger programs had converged, the calculated products were compared with the specifications and the temperature profiles in the columns examined. At a given feed pressure, the heat load distributions were adjusted to give maximum separation in each column and the calculations repeated. The degree of separation was assessed very easily from the temperature profiles. Throughout these calculations, the mixtures behaved effectively as binary systems since no product stream was of sufficiently high purity to cause argon to accumulate within the columns. When a configuration of heat loads had been found which gave satisfactory temperature profiles, attention was turned to the product compositions. The heat loads were increased or decreased to change the internal vapour-liquid ratios in the columns. Lastly the calculated temperature differences at each link were compared with the specified heat loads. Excessive variations in the ratios of heat loads to temperature differences were smoothed out by redistribution of loads, while the air feed pressure was modified to maintain feasible surface areas in the various condenser-reboiler links.

Although this strategy lacks sophistication and seems to eschew modern optimization techniques, the study is concerned with only one feed stream of fixed composition and many other process variables are constrained by external factors, so solutions were found quite quickly.

Since extensive simulations are required to establish detailed configurations, it is necessary to set up good initial systems as a basis for evolution of improved flowsheets. The distribution of thermal links serves to create more reversible separations within the columns as well as maximise the temperature differences at the links at a given operating pressure. Consequently one starting point is to consider a fractionating condenser staged system for the HP column in which the internal vapour and liquid flows are adjusted stage by stage to maintain the overall separation at a given total number of stages. The calculations give a heat load profile which emphasises the possibilities of maintaining high condensing side temperatures. A practical configuration using external condensers (6) enables even more favourable distributions to be obtained since the effective condensing temperatures can approach the dewpoints of the vapour sidestreams and the returning condensate streams can be passed to stages higher in the column whose phase compositions match more closely. Calculations over a wide range of conditions show that about half the total condensing load is liberated from the overhead product vapour and about half from the sidestreams taken over the full height of the column even when internal flows are maintained to give nearly reversible separation.

In a similar way, the LP column can be approximated by a distributed system in which the smallest possible reboil loads are used near the bottom still maintaining separation at each stage. In this case, the total heat load and temperature distribution vary widely with oxygen product purity and recovery. The most serious disadvantage of the rating approach is its inherent lack of ability to fix the oxygen product purity and so identify a heat load distribution for the LP column. Since the oxygen product is taken from the column as a liquid the entire heat load to the LP column is distributed so the temperature profile for the thermal links does not show the flat region corresponding to the condensation of the overhead vapours from the HP column. This implies that the smallest temperature differences between the condensing vapour sidestreams and the evaporating liquid in the thermal links will be found near the bottom of the LP column. Once profiles of heat load versus stage number have been estimated from these simple models, the profiles can be approximated by an equal number of loads placed in each column to give a feasible initial set of temperature differences at each link.

Calculation of the distillation columns

The column calculations used extensions of the well known bubblepoint method (8). In the case of the HP column, the model contains many vapour sidestreams with return of the condensates to the column, or to other condensers, or to a collecting point from which a mixed overhead product goes to the LP column. For a given set of profiles of temperature, internal liquid and vapour flows, the profiles of composition on each stage were found by solving the usual trdiagonal matrix equation, component by component. The temperatures, enthalpies and component K-values were estimated from the normalised liquid compositions. The profiles of internal vapour and liquid flowrates were then calculated using the enthalpy data just found. This enthalpy balance which also included the external condensers, gave new estimates of the return streams from the condensers, i.e. the feed streams used in the composition calculations. The calculations were then repeated until satisfactory convergence was obtained on composition, temperature and internal flowrate profiles. The external condensers were modelled in several ways depending on the purpose of the calculation. In the simplest model, the condensing heat loads and the vapour fractions of the streams returning to the column were fixed. The calculation then gave results for the flowrate of the vapour sidestreams and the effective temperature of the condensing mixture. In other cases, the flowrate of the vapour sidestreams, the overall heat transfer coefficient, the surface area of the link and the effective evaporating liquid temperature were specified giving results for the extent of condensation and the effective condensing side temperature. Factors such as distillation stage pressure drop, stage efficiency, heat leak into each stage, pressure losses in the external condensers were included in the model.

Calculation of the LP column was more straightforward. Although the absence of an overhead condenser, the multiple feeds and reboilers made the configuration different from the simple column of the textbooks,the bubblepoint algorithm required no modification and gave reliable results. The heat loads at the reboilers could be modelled as set loads or as specified overall heat transfer coefficients, surface areas and effective condensing side temperatures. The distillation column programs proved reliable and always gave physically realistic results even when the internal vapour or liquid flows became zero. All calculations converged to 0.001 K and mole fractions were reproduced to the fourth decimal place. The program to simulate the heat exchanger system used simple performance models of the equipment in which specified approach temperatures and data describing the input streams were used to calculate the conditions of the output streams, heat loads, etc. Several different configurations of equipment were studied to devise schemes with improved energy recovery and distributions of the recycled cold which were most advantageous to the separation requirements of the system.

Data for the simulations

Physical property data for nitrogen-argon-oxygen mixtures are available from many sources (3). The correlations used to calculate vapour liquid equilibria were those of Harmens (4). These agreed very well with other information reviewed by Hempseed (7) and were in a particularly convenient form for the simulation calculations. Some column calculations have used a stage to stage approach (1) in which the primary variables are the compositions of the co-existing phases. In this work, the heat transfer aspects are very important, also the chosen calculation method is a modification of the bubblepoint technique. Consequently, temperature is the primary variable. Since Harmens' equations are explicit in temperature, they are more suitable than those of Latimer (9). Enthalpy data were reviewed by Hempseed (7) and the correlations shown in Table 1 derived to represent the available data with adequate accuracy in the particular range of temperature, pressure and composition of interest.

The other data required describe the performance of the equipment. The pressure drops across the distillation trays were assumed in the range 0.003 to 0.007 bars for the HP column and 0.007 to 0.010 bars for the LP column (2). As discussed later, the values for the LP column are more critical and less easily estimated than those for the HP column. The success of the proposed process depends to a large extent on reducing the pressure above the evaporating froths in the LP column, and on accurate estimation of the composition and equilibrium temperatures of the phases. This requires knowledge of the stage efficiencies. Values were assumed as 100% and 70 to 100% for the HP and LP columns respectively. Other data required are the heat transfer coefficients between the condensing vapour from the HP column and the froth on the trays in the LP column. Literature search showed a serious lack of data, particularly for the contribution of the boiling side. For calculation purposes, values in the range 1 to 3 kW. $m^{-2}.K^{-1}$ were used for overall temperature differences in the range 2 to 5 K. It was anticipated that the froth and vapour flow through the froth would increase these values severalfold. Although the effect of composition in the range 40 to 95% O_2 liquid was reviewed, in view of the other uncertainties, composition effects were ignored for both heat transfer coefficients and efficiencies.

Experimental Investigations

To obtain preliminary confirmation of the assumed values of stage efficiency, pressure drop and heat transfer coefficients, a mixture of chlorinated paraffins was distilled at atmospheric pressure in a small test rig (7). Vapour flow through a small sieve tray was maintained by a closed loop circulation from a reboiler to a total condenser with the liquid condensate refluxed to the tray or bypassed to the reboiler. An independent reboiler-condenser loop supplied vapour at a higher pressure to a single heat transfer baffle immersed in the froth created on the seive tray. Different compositions and pressures of the vapour condensing in the baffle were chosen to simulate different conditions within the LP column.

Three different situations could be distinguished within the LP column. At the base, the heat transfer conditions correspond to a combination of natural convection and pool boiling since the distilling vapour flowrates and baffle temperature differences will be small. The low flowrate of distilling vapour eases the conflict between the requirements of low tray pressure drop, i.e. vapour flow area and high heat transfer rates, i.e. large baffle surface areas and probable reduction in distilling vapour flow area. Further up the LP column, simulation indicates that the baffle temperature difference remains small even though the distilling vapour flowrate increases. If the product purity exceeds 90% the baffle temperature difference will probably decrease slightly relative to the value at the column base. Finally, in the uppermost links, the vapour flowrates will be large and the baffle temperature differences will increase. While the heat transfer problem may ease, the requirement of low stage pressure drop and high efficiency remains. The experimental work is reviewed by Hempseed (7). Although further work is required to confirm the results, especially the extrapolations to low temperatures and air mixtures, the estimates from the literature were confirmed. One factor not covered by the present work is the use of promoted surfaces (10) and their effect not just on heat transfer but also on mass transfer within the distillation froth.

Simulation Results

Very good approximations to the ideal of a continuously varying reboil and reflux distributions can be obtained using about 15 theoretical stages in the HP column and between 25 and 35 theoretical stages in the LP column (most of the variation in the LP column arises above the feeds in attempts to obtain high oxygen recoveries.) Although both columns work near to pinched conditions, the performance of the HP column is relatively insensitive to small changes in the distribution and location of the heat links. Consequently the main effort of the simulation search was devoted to obtaining good, i.e. uniform, separation performance from all stages in the LP column below the feeds (7)(11). The most effective limit to the approach to reversible working was the strong interaction between the purity of the oxygen product, the recovery of oxygen and the condition of the liquefied air feed entering the LP column from the oxygen product evaporator. Any deterioration in the liquid reflux caused lower product purity and lower oxygen recovery, which in turn led to a change in the performance of the evaporator closing the loop affecting the liquid reflux.

A number of cases were studied in which different heat link configurations were used while attempting to steer the designs towards targets of 80% to 95% oxygen purity. The results of most interest include the feed air pressure and Figure 3 shows the band within which all results lie. If these data are compared with the pressures required in a single link process, energy savings of from 15% to 25% should be achieved. The configurations differed mainly in the number of thermal links used. For a particular oxygen purity, the feed air pressure increased as the number of links decreased. If the number of links was reduced below six, the feed air pressure increased markedly while configurations using ten or more links showed only slight decreases in feed air pressure. From a practical point of view, the cases using more than ten links are of interest as pointers to the potential of the concept once suitable designs of tray incorporating heat transfer surfaces have been developed. On the other hand, the studies using fewer links are of interest as interim proposals which could be implemented using existing equipment. The single condenser-reboiler used in existing plant often consists of a number of extended surface exchanger blocks arranged in an annular fashion around the column shell. In a large capacity plant using a few large heat links, the duties might be met by placing similar blocks around the LP column shell at appropriate levels. Some results are shown in Figures 4 and 5 for a system using eight links to give a 94% purity oxygen product using a feed air pressure of only 4.7 bar.

Conclusions

The use of distributed reflux and reboil to achieve energy savings has been recognised in the past. The present work describes simulations of detailed performance models to provide quantitative evidence of the benefits and point to areas requiring future development. In the long term, improved distillation trays with integral heat transfer elements will provide opportunities to implement the multiple heat link idea to its full advantage. In the short term, some advantage can be obtained by using existing equipment, i.e. extended surface exchanger blocks in designs using a few links of large size. In such a case, energy savings of at least 10% are found in designs producing a 95% purity oxygen product.

Acknowledgements

We gratefully acknowledge the support of Cryoplants Ltd for the research program of J.W.Hempseed, also for the assistance provided for A.C. Powell during his M.Sc. project.

LITERATURE CITED

1. Armstrong M. and A.E. Schofield, The Chemical Engineer (London) p.CE 184, IChemE. May (1969).
2. Connelly, K.E. and G.G. Haselden, 'Distillation', 3rd International Symposium, IChemE. (London) p.2.3-65 (1979).
3. Din, F., 'Thermodynamic Functions of Gases', Butterworths (London), vol.2, 1, (1962).
4. Harmens, A., Cryogenics, 10, 406, (1970).
5. Haselden, G.G., Trans IChemE. (London) 36, 8 (1958)
6. Haselden, G.G., Brit.Pat.Application 16391/74.
7. Hempseed, J.W., PhD. thesis, Univ. of Leeds, (to be submitted).
8. King, C.J., 'Separation Processes' J.Wiley, NY. (1980).
9. Latimer, R.E., AIChE Journal, 3, 75, (1957).
10. O'Neill, P.S. and C.F. Gottzmann, The Cryogenic Processes and Equipment Conference, Century 2 - Emerging Technology Conference, p.37., San Francisco, August 1980, ASME.
11. Powell, A.C., M.Sc. dissertation. University of Leeds, (1981).

Table 1

Liquid Phase:

Nitrogen $H_l = 5716.6+T(103.68-T(0.6346-0.002898T)))$

Argon $H_l = -3532.0+44.15T$

Oxygen $H_l = -4000.2+T(46.542+0.04325T))$

Vapour Phase:

Nitrogen $H_V = 3268.6+T(29.1305-0.001648T)+P(-392.81+T(5.697-0.0224T)))$

Argon $H_V = 5061.7+20.75T+P(-123.42+0.44T))$

Oxygen $H_V = 4777.0+T(30.234-0.00731T)+P(-357.73+T(4.4169-0.01457T)))$

Basis is zero enthalpy for pure liquids at 80 K

Units are K, bars and $kJ.kmol^{-1}$

Range of use is 0-5% Argon, 1 - 4.5 bars.

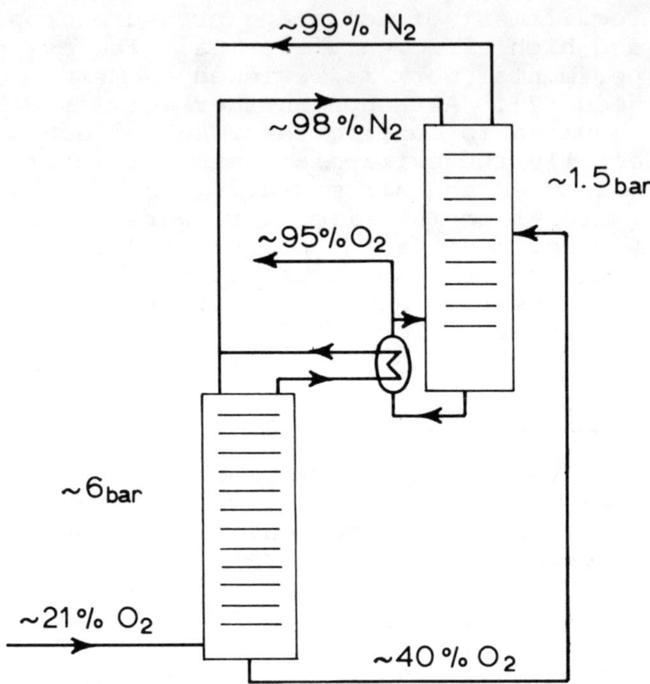

Figure 1. Air separation by low temperature distillation.

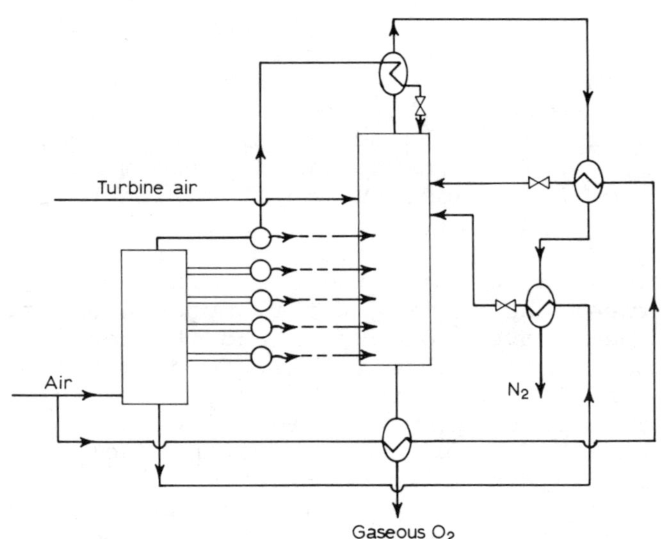

Figure 2. Distributed reflux and reboil.

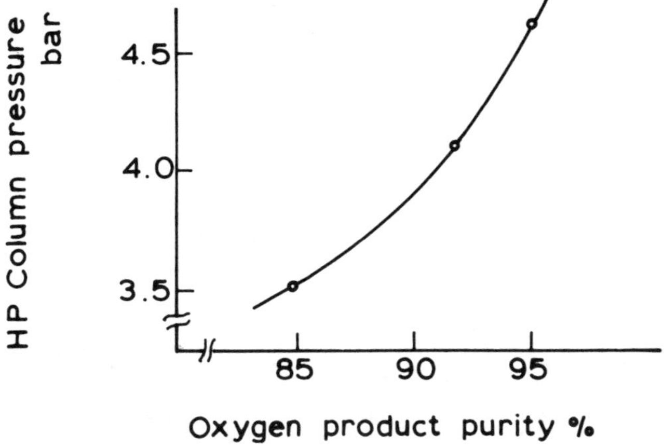

Figure 3. HP column pressure versus oxygen product purity.

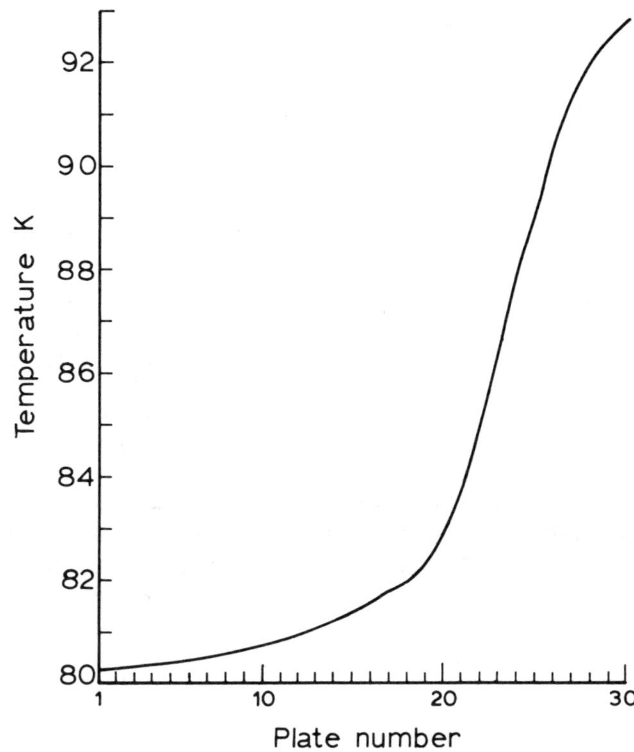

Figure 4. LP column temperature profile.

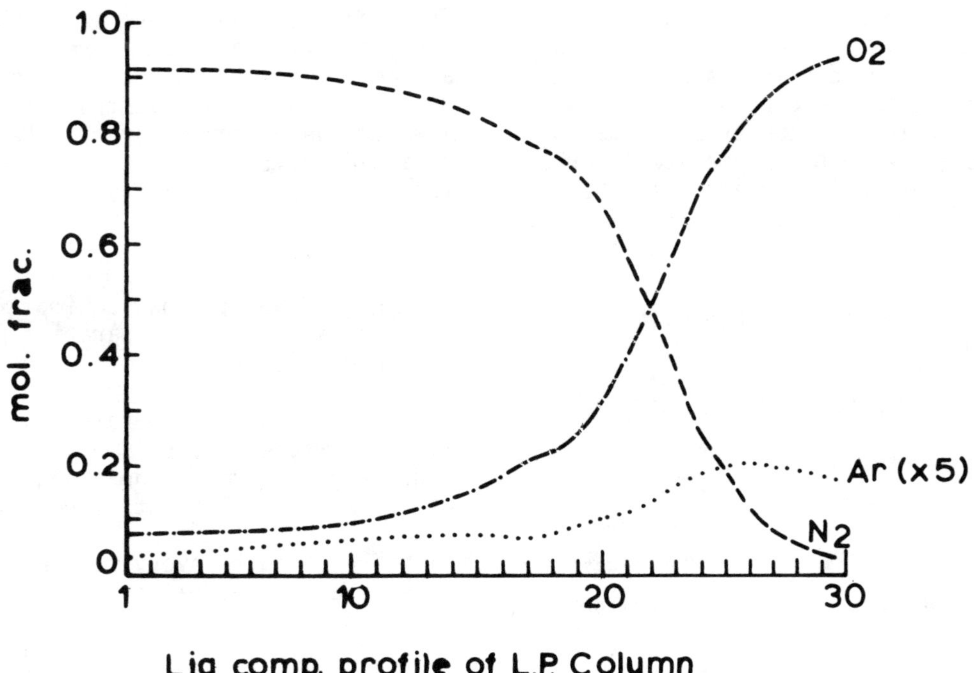

Figure 5. LP column composition profiles.

METHODS FOR ARGON RECOVERY TO MEET INCREASED DEMAND ON THE ARGON MARKET

HELMUT SPRINGMANN

LOTEPRO CORPORATION
1140 Avenue of the Americas
New York, N.Y. 10036

This paper presents methods for economic production of argon. The growing market for shielding gases, especially argon, and especially in the steel industry, requires optimal processing. The market characteristics, basic processes not only from air but from ammonia and other sources, and comparative process economics are considered.

In the past years, there has been a tremendous increase in argon consumption. This increase has been primarily caused by the demands of the steel industry.

Argon has several applications in steel works. It can serve, for example, as a protection medium in the continuous casting machine, to decrease the carbon monoxide partial pressure in the AOD-converter, or as a shielding gas for welding. Some steel works have recently started to use argon in the BOF converter. For this purpose, roughly one Nm^3 (35 SCF) is consumed for one ton of steel.

Table No. 1 shows the argon balance of a normal steel works.

TABLE 1: ARGON RECOVERY IN A STEEL WORKS.

STEEL PRODUCTION	=	3 Mio tons/year 375 tons/hour
OPERATING HOURS		8000 hours/year
OXYGEN CONSUMPTION		60 Nm^3/ton steel
OXYGEN PRODUCTION	=	22 500 Nm^3/hour 850 STD
AIR CONSUMPTION	=	112 500 Nm^3/hour 69 780 SCFH
ARGON RECOVERY		50 vol %
ARGON PRODUCTION	=	524 Nm^3/hour 25 STD

ARGON RECOVERED FROM AN AIR SEPARATION PLANT

It is assumed in this example of a medium sized steel works that there is an air separation plant for production of oxygen. The argon is also recovered in the cryogenic air separation plant. There is only as much air compressed as is required for oxygen production. There is no extra air for the argon. The oxygen recovery rate is kept high to avoid additional costs. The expenditure for the crude argon is only for the slightly increased air pressure due to the higher number of trays.

Figure No. 1 shows a simplified flowsheet of a large air separation plant, a low pressure facility.

The argon is withdrawn from the middle of the low pressure column and sent to the crude argon column. The argon concentration at the withdrawal point is roughly 10%.

The argon recovery rate at this point depends on two conditions:

1. The purity of the product oxygen, and
2. The purity of the waste nitrogen

The purity of the waste nitrogen should be high for recovering most of the oxygen. In case of a plant with a given number of trays, the purity of the oxygen can be adjusted by the flowrate of the oxygen, while the nitrogen purity must be adjusted by the flowrate of the expansion turbine. Both adjustments affect the reflux ratio, which obviously determines the purities, besides the number of trays.

Figure No. 2 shows the argon losses with the product oxygen flow as a function of the oxygen concentration in the product. In practice, the tailgas in the product oxygen is argon due to the small difference in vapor pressure between argon and oxygen compared to argon and nitrogen.

Figure No. 3 shows the argon losses in the waste nitrogen as a function of the amount of injection air. It is assumed that high purity oxygen (99.7% O_2) is produced and the maximum possible amount of argon is recovered. An optimum number of trays is installed.

If a high argon recovery rate is to be obtained, the air injection must be reduced or shut off. Pressurized N_2, which is taken from the medium pressure column, has the same result as injection air. This means that power must be expended if the argon recovery rate is to be increased. The power is required for compressing additional turbine air or nitrogen.

Figure No. 4 shows the amount of specific power which must be spent for the production of crude argon.

At an argon recovery rate of 30%, no additional power is required in an air separation plant, which produces 99.5 to 99.7% oxygen. Below 30% there is even a power bonus because of the removal of argon, which blocks the rectification process.

At higher recovery rates, the number of trays increases, which thus increases the pressure drop, and so that energy must be spent for the production of cold. The specific power consumption for the argon production increases with the increasing argon recovery rate. As stated earlier, additional power is required for the purification and liquefaction of the argon

ARGON RECOVERY FROM AMMONIA PLANT PURGE GAS

In the following, the power consumption of argon recovery with other processes is shown. Figure No. 5 shows the block diagram of an ammonia factory. Generally, a lower O_2 purity is required than in a steel works.

Within the air separation plant it is advisable to decrease the oxygen purity to 95% O_2. Between 80 and 90% of the argon contained in the air can be drawn with the oxygen. The argon-containing oxygen stream is sent to the gasifier, shift converter, CO_2 removal, and finally to the nitrogen wash unit. In the nitrogen wash unit, the argon is washed down with carbon monoxide and methane. A tailgas fraction is formed and sent to a separation unit.

In a cryogenic unit three products are obtained. The largest fraction is stochiometric hydrogen-nitrogen mixture for the ammonia plant. There is a small fuel gas fraction and finally a liquid argon fraction.

Figure No. 6 shows the separation unit, which contains the separation columns. There is the nitrogen wash column (No. 3), the methane removal column (No. 4) and the argon column (No. 5). The cold is supplied by a nitrogen cycle.

For calculation the power consumption of the additional separation units, it can be assumed that the heating value of the combustibles is the same as that of the raw materials. The same amount of steam must be spent. Additional mechanical power is only required for the compression of the gas and of the nitrogen. This electrical power can be equally distributed to all three products. There is no preference for one product. The gaseous argon product therefore must be charged with 0.2 kWh/Nm^3. This is the only power charged to the argon. It must be recalled that thre is a high argon recovery rate of 80%. This indicates that the argon made in an optimized ammonia factory requires less power for high recovery rates, as compared to the argon recovery within the air separation unit. (See Figure 4).

If synthesis purge gas is available, the power consumption can be even lower. Synthesis purge gas is available in an ammonia plant based on the steam reforming process. The argon comes with the air, the nitrogen of which is required, and which is introduced via the secondary reformer. The argon does not react and is concentrated in the high pressure synthesis tube. The argon must be released and thisis done with the purge gas. The purge gas has the following composition:

TABLE 2: PURGE GAS COMPOSITION

H_2	67 vol %
N_2	22 vol %
Ar	3 vol %
CH_4	8 vol %

The purge gas is available at high pressure. Therefore no additional compression of the purge gas is required.

The specific power for the production of gaseous argon is 0.1 kWh/Nm^3. The calculation was done the same way as described in the argon recovery of nitrogen scrubbing tail gas.

ARGON RECOVERED FROM THE AOD WASTE GAS

One of the major argon consumers is the AOD process. 16-25 Nm^3 Ar per ton of steel are required in the AOD converter. It is obvious that the recycling of the argon contained in the converter waste gas can be of interest.

The waste gas of an AOD converter contains the following constituants:

TABLE 3: CONVERTER WASTE GAS COMPOSITION

H_2	0.8 vol %
N_2	20.6 vol %
CO	42.8 vol %
Ar	35.8 vol %

Figure No. 7 shows the flowsheet. The crude gas comes from the AOD converter at ambient pressure. It is sent to a water washing system for cooling and the removal of dust. Because of the intermittent operation of the converter, the waste gas must be collected in a wet gas holder. It is then compressed by a steadily working raw gas compressor to the pressure required in the separation column. The raw gas is cleaned, and carbon dioxide and water gas are removed. In the separation column, the argon is separated from the H_2-N_2-CO fraction and is concentrated in the bottom of the column. The energy of the liquid bottom product is used for producing the reflux in the column by aid of a condenser.

In the case described, additional reflux is produced by a raw gas recycle, which is economical in special cases, but which is not always. The power consumption is 0.75 kWh/Nm^3 of argon. This indicates that the argon recovery of an AOD converter waste gas is not the most economical process. It is economical insofar as it can be used if no other argon is avaiable.

COST COMPARISON OF THESE PROCESSES

Table 4 shows the specific depreciation costs for 3 processes.

TABLE 4: SPECIFIC DEPRECIATION COSTS OF ARGON

PROCESS	AIR SEPARATION	N_2 WASH TAILGAS SEPARATION & SYNTHESIS PURGE GAS SEP.	AOD WASTE GAS SEPARATION
PRODUCTS	Ar	Ar $3H_2 + H_2$	Ar
SPECIFIC PRODUCTION COSTS (Financial units per Nm^3)	0.09	0.045	0.16

The specific depreciation costs are calculated on the basis of 15% depreciation and 8000 operating hours per year. The depreciation costs are within economical limits.

If power and depreciation costs of all three processes are compared, the figures of the fourth process, synthesis purgegas separation, are similar to the N_2-wash tailgas separation, it can be said that all the processes are economical. The N_2-wash separation, however, is the most economical system. This is for high recovery rates. For low recovery rates, the air separation process may be better.

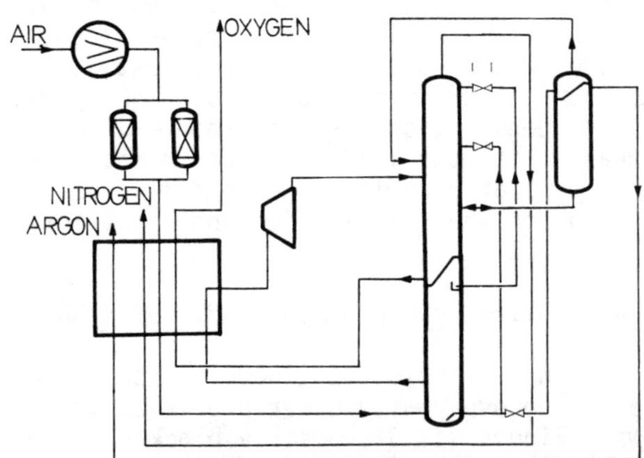

FIG.1: AIR SEPARATION PLANT WITH CRUDE ARGON COLUMN

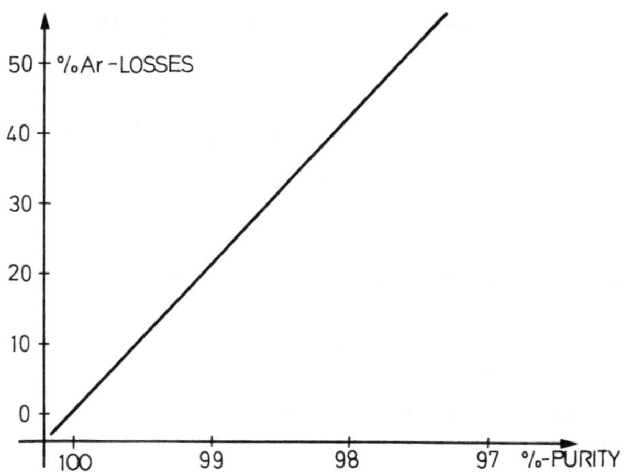

FIG.2: ARGONLOSSES WITH THE OXYGEN PRODUCT STREAM

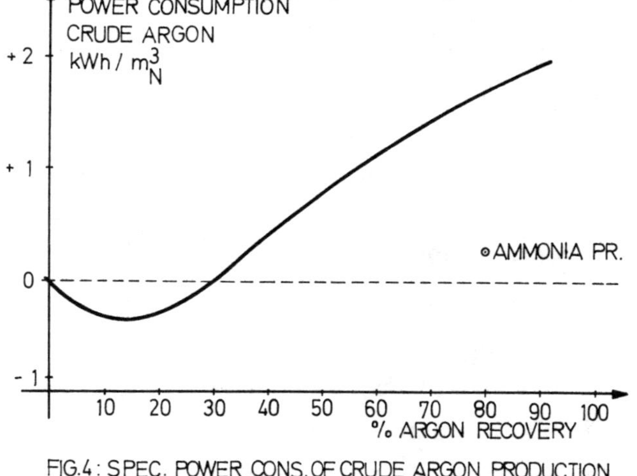

FIG.4: SPEC. POWER CONS. OF CRUDE ARGON PRODUCTION

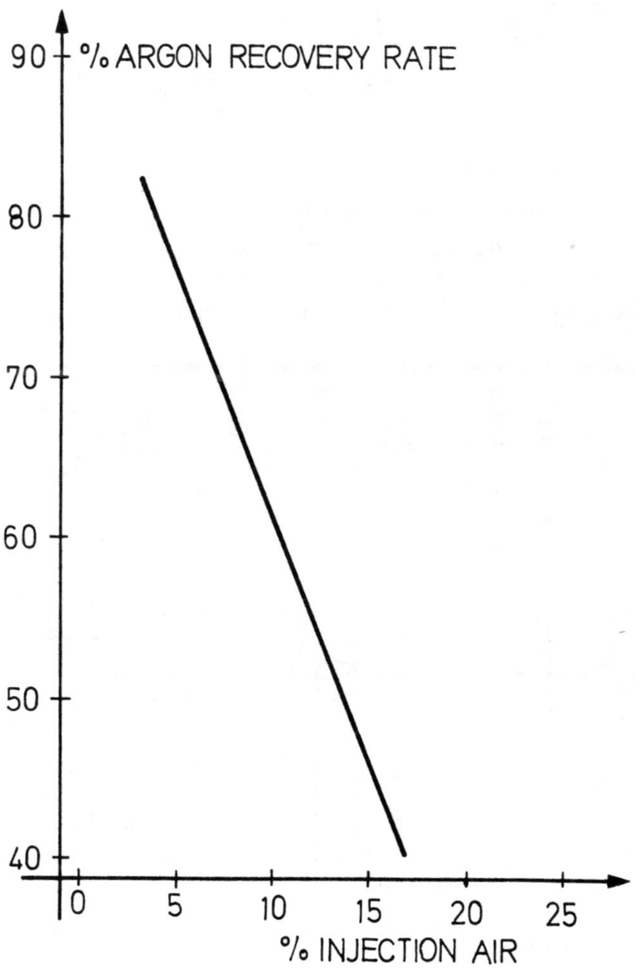

FIG.3: ARGON RECOVERY RATE

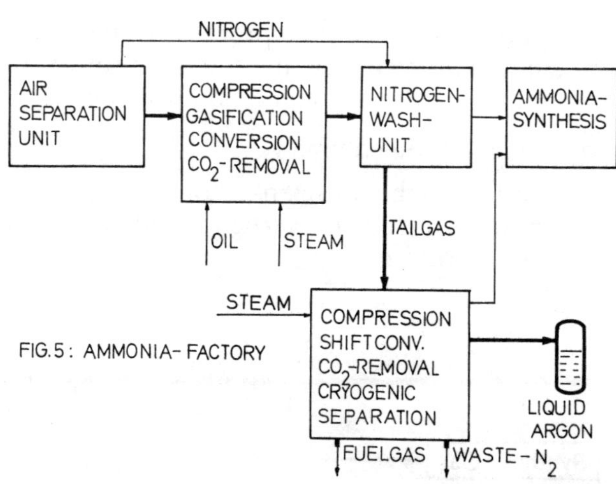

FIG.5: AMMONIA-FACTORY

Gas purification

1 Adsorber

Removal of water and ammonia in interchangeable adsorbers to prevent freezing in the cryogenic section

Gas separation

2 Heat exchanger

Cooling of feed gas
Reheating of separation products

3 Wash column

Removal of argon and methane by scrubbing with liquid nitrogen

4 Fractionation column

Splitting of the bottom product from wash column 3 into a methane fraction (bottom) and a nitrogen-argon fraction (overhead)

5 Fractionation column

Splitting of the nitrogen-argon fraction into synthesis nitrogen (overhead) and pure argon (bottom)

6 Tank

Storage of liquid nitrogen
Cooling of wash nitrogen and synthesis nitrogen in a heat exchanger coil

7 Valve

Feeding of synthesis nitrogen into the H_2+N_2 mixture

8 Compressor

Joint compression of recycle and synthesis nitrogen

9,10 Heat exchangers

Cooling of high-pressure nitrogen
Reheating of low pressure nitrogen

11 Expander

Expansion of part of the high pressure nitrogen for the generation of refrigeration

Energy balance

A high-pressure nitrogen cycle with expander generates the refrigeration required for argon liquefaction and for compensation for heat leakage and non-ideal heat exchange

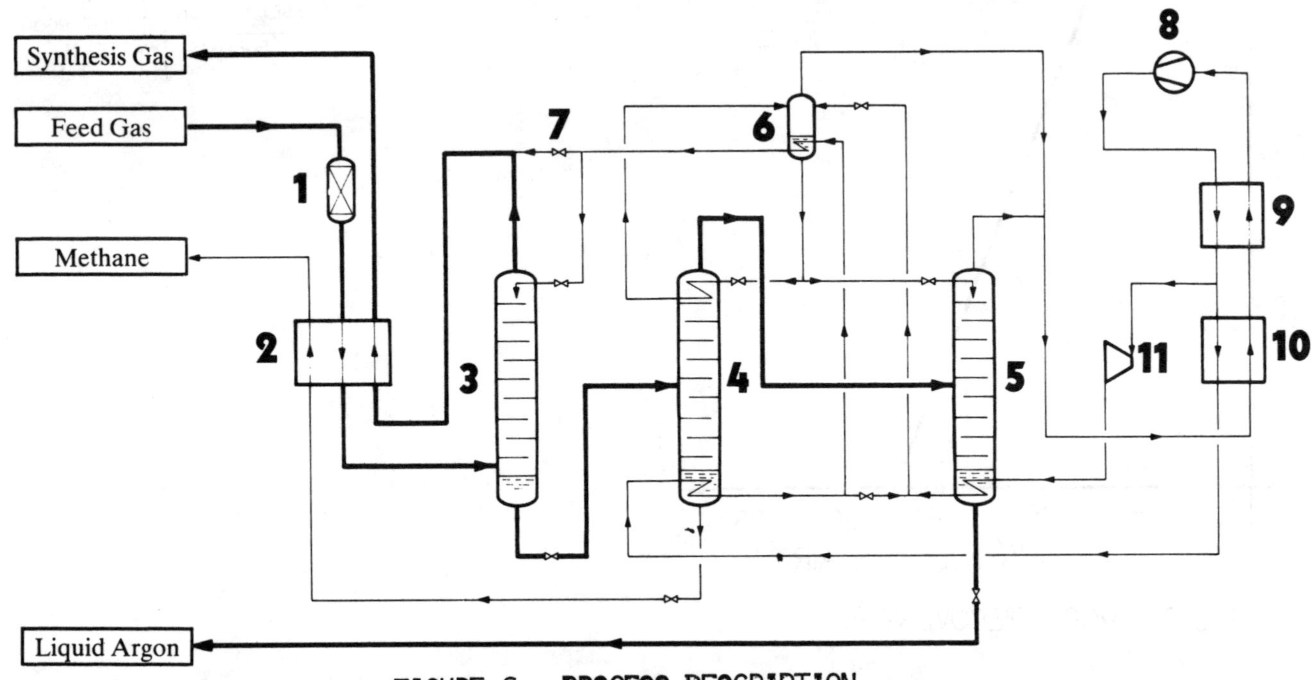

FIGURE 6: PROCESS DESCRIPTION

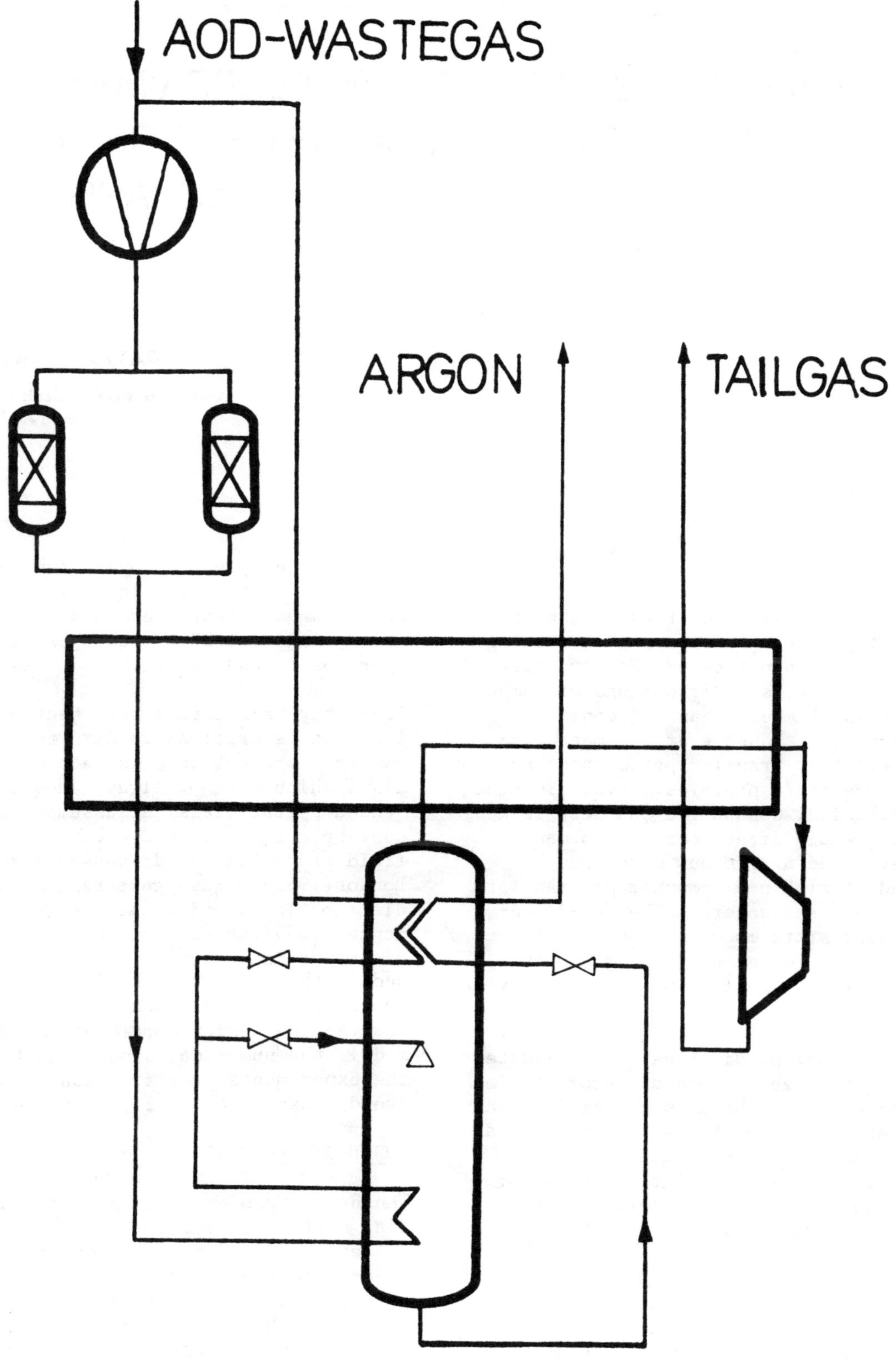

FIG. 7 : ARGON FROM AOD-CONVERTER WASTE-GAS

COMPUTER SIMULATION AND EXPERIMENTAL VERIFICATION OF THE PERFORMANCE OF CORE HEAT EXCHANGERS

KENNETH J. ANSELMO
and
DAVID J. HERSH
Air Products & Chemicals Inc.
Allentown, PA

SUMMARY

Core heat exchangers, ubiquitous in the cryogenic industry, present complex heat exchange mechanisms which cannot be adequately analyzed using the classical simplifications of lumped cooling curves, lumped areas and constant fluid properties. Passage arrangement controls overall heat transfer performance and internal temperature profiles. Flow distribution, longitudinal conduction and heat leak also significantly affect core performance. We have developed a rigorous model and implemented it within a computer program for simulating core exchangers. The one-dimensional, steady-state model accounts for heat transfer both by forced convection and by conduction through the fins on a passage by passage basis.

The detailed model predictions were validated by measuring passage temperature profiles for several in-service exchangers. Overall checks on outlet stream temperatures for many operating plants have given us added confidence in the model results. All Air Products core heat exchangers are analyzed through simulation prior to final design.

INTRODUCTION

Core heat exchangers are an integral part of many industrial cryogenic plants. Combining low mass with high heat transfer efficiency, these exchangers are ideal for many cryogenic heat transfer applications: air separation, nitrogen rejection, nitrogen wash, ammonia purge gas recovery, hydrogen purification, hydrogen-carbon monoxide separation, helium recovery, ethylene recovery and carbon dioxide recovery. During the last five years Air Products has spent over 50 million dollars in purchasing core heat exchangers and currently operates thousands of these cores worldwide.

Obviously, simulating core heat exchanger performance is critical to Air Products. Cores, however, present complex heat exchange mechanisms which are not always adequately represented by the classical assumptions of lumped cooling curves, lumped areas and constant fluid properties as discussed by Kays and London(1955). Passage arrangement and flow distribution significantly affect core performance; longitudinal conduction and heat leak are often important. These effects are included in the model used by Air Products and in the corresponding computer module used to analyze and design core heat exchangers. The model, the numerical simulation technique and the experimental verification of the model are discussed in the following sections.

MODEL DEVELOPMENT

The heat transfer section of a core (plate-fin) heat exchanger consists of parallel plates separated by fins connected perpendicularly to these plates. The fluid flows parallel to the plates plates along the length of the fins. Figure 1 shows the assembly for the parallel flow pattern which we consider while Figure 2 depicts the general single layer structure.

The basic assumptions in the model development are:

1. Energy and momentum balances can be solved separately.

2. Thermodynamic equilibrium exists between vapor and liquid phases throughout the two-phase region.

3. Temperature variations across the core width are neglected.

4. Temperature drop across parting sheets is neglible.

5. Fin perforation or serration effects are ignored when calculating fin temperature profiles.

6. Longitudinal conduction effects are ignored.

7. Heat leaks are ignored.

Several points concerning these assumptions are in order. First, since pressure drop variations are small the heat transfer calculations may be based on tables of thermodynamic and transport properties. This eliminates the very time consuming calls to a thermodynamic package and allows us to decouple the energy and momentum equations. In this paper only the energy equations are considered. Second, non-equilibrium effects are included by correction factors applied to the heat transfer coefficients. Finally, although longitudinal conduction in the parting sheets and heat leak from the cap sheets have been neglected in our formulation, their inclusion in the model and in the solution algorithm is straightforward. These latter effects become increasingly important for very cold exchangers with low mass flow rates and correspondingly small heat fluxes.

The equations describing the heat trans-in and plate-fin exchanger consistent with the above assumptions may be found in Weimer (1972) and Kao (1961). We present them here in a form suitable for describing the numerical algorithm to be proposed in the next section. Figure 3 shows an idealized passage cross section.

An energy balance on the i'th parting sheet coupled with the the fin temperature profiles and the assumption of uniform parting sheet temperature yields:

$$\left[\frac{-\gamma_{i-1}}{s_{i-1}}\right]TP_{i-1} + \left[\frac{\gamma_{i-1}}{s_{i-1}}c_{i-1} + \beta_{i-1} + \beta_i + \frac{\gamma_i}{s_i}\right]TP_i + \left[\frac{-\gamma_i}{s_i}\right]TP_{i+1} =$$

$$\left[\frac{\gamma_{i-1}}{s_{i-1}}(c_{i-1} - 1) + \beta_{i-1}\right]T_{i-1} + \left[\frac{\gamma_i}{s_i}(c_i - 1) + \beta_i\right]T_i \qquad (1)$$

$$i=1,\ldots,n+1$$

where $\beta_0 = \gamma_0 = \beta_{n+1} = \gamma_{n+1} = 0$

and

$$\alpha_i = \left[\frac{24h_i}{kt_i}\right]^{1/2}$$

$$s_i = \sinh(\alpha_i f_i/12)$$

$$c_i = \cosh(\alpha_i f_i/12)$$

$$\beta_i = h_i AP_i$$

$$\gamma_i = kAF_i\alpha_i$$

$$AP_i = W(1 - t_i d_i)$$

$$AF_i = Wt_i d_i$$

An energy balance on the fluid in the i'th passage yields:

$$\frac{d\hat{H}_i}{dz} = (TP_i + TP_{i+1} - 2T_i)\left[\beta_i + \frac{(c_i-1)}{s_i}\gamma_i\right] \Big/ m_i \qquad (2)$$

$$i=2,\ldots,n-1$$

where m is the molar flow rate. It is a signed number where the sign denotes flow direction. Flows are usually uniformly distributed but by including passage molar flows in the model flow maldistribution may be studied.

Inlet stream temperature, pressure and composition give the initial conditions for equations (2). Observe that since counter flow is allowed a two-point boundary-value problem (TPBVP) arises. That is, initial conditions for the differential equations (2) are prescribed at the point where the stream enters the core. Unfortunately, unlike an initial value problem where all initial conditions are prescribed at one spatial point and approximate solutions may be calculated by marching in the independent variable, solving a general TPBVP is an iterative procedure. We discuss a very stable technique in the next section.

signed number where the sign denotes flow direction. Flows are usually uniformly distributed but by including passage molar flows in the model flow maldistribution may be studied.

Inlet stream temperature, pressure and composition give the initial conditions for equations (2). Observe that since counter flow is allowed a two-point boundary-value problem (TPBVP) arises. That is, initial conditions for the differential equations (2) are prescribed at the point where the stream enters the core. Unfortunately, unlike an initial value problem where all initial conditions are prescribed at one spatial point and approximate solutions may be calculated by marching in the independent variable, solving a general TPBVP is an iterative procedure. We discuss a very stable technique in the next section.

NUMERICAL ALGORITHM

The solution technique described in this section has been very stable for all Air Products applications. In a general setting it is called successive approximations and when applied to differential equations is named Picard iteration. It is usually seen when proving existence-uniqueness theorems for initial-value problems. Before applying this to our TPBVP a shooting method was tried but it was unstable for even the simplest case. The successive approximation method is presented here as applied to a single equation; extension to a set of equations is straightforward. Further discussion of the method may be found in Braun (1975).

Consider the problem

$$\frac{dy}{dz} = g(y), \quad y(a) = b. \tag{3}$$

The successive approximation method generates a sequence of functions y(z;1), y(z;2)... which converges to the solution of (3). That is, given y(z;j) we calculate

$$y(z;j+1) = y(a) + \int_a^z g(y(s;j))\, ds \tag{4}$$

Defining $r(y;j) = \frac{d(y(z;j))}{dz} - g(y(z;j))$

and inserting a damping factor, p, to promote stability (4) may be rewritten:

$$y(z;j+1) = y(z;j) - p \int_a^z r(y(s;j))\, ds \tag{5}$$

Note that the residual, r(y;j), is the difference between the actual derivative of the j'th iterate and the derivative calculated from the model equations; that is, the right hand side of (2). When these are equal the model is satisfied: r=0 and the correction (integral) term in (5) is identically zero.

Equation (5) is our basic iteration formula. In order to use (5) to solve model equations (1) and (2) a grid of equally spaced points over the exchanger length is defined and an approximate solution at these points is sought. The equal spacing is for numerical convenience; a grid spacing of about 3 inches is adequate for all our problems. The first step in obtaining a solution is to guess the fluid enthalpy profiles, H, in each passage at each grid point. In practice this is done by setting up linear profiles based on inlet stream conditions and a user-supplied temperature difference estimate. Our experience shows that these initial profiles affect the convergence rate but usually not convergence itself. A damping factor, p, of 0.1 is the default value although this is dynamically adjusted as the calculation proceeds.

Having assumed H profiles (and hence the fluid temperatures, T) the corresponding parting sheet temperatures, TP, are obtained by solving (1). These are a set of tridiagonal linear equations in TP since the right hand side depends only on the guessed T profiles. Furthermore, the TP at a fixed z are independent of the TP at any other z. This makes the TP calculation a very easy calculation for any set of fluid temperatures. When longitudinal conduction in the parting sheets is considered, this situation no longer holds but we found that an additional iteration which converges very quickly and retains independent calculation of TP at planes of constant z could be incorporate into the calculation scheme.

Once the parting sheet temperatures, TP, are known these may be used along with the assumed fluid temperatures to calculate the residual functions needed in applying (5) to calculate new fluid molar enthalpy profiles. dH/dz (the left hand side of (2)) is calculated using a 5-point difference scheme (based on interpolating polynomials) on H, r is formed by subtracting this from the right hand side of (2) and Simpson's rule is used to perform the required numerical quadrature. These numerical techniques are discussed in Conte and de Boor(1980); working equations are given in Perry's Handbook (1963). The TPBVP is automatically handled since for each passage (5) is integrated beginning at the point where the stream enters the exchanger. That is, where its initial condition is known. Once we have new profiles the process is repeated until the undamped fluid profiles change less than prescribed tolerances.

A summary of the numerical procedure follows:

1. Guess fluid profiles, H.
2. Calculate parting sheet temperatures, TP, from (1).
3. Calculate residual functions using (2) and numerically differentiating current H.
4. Use (5) to calculate new profiles.
5. If undamped profiles have changed less than prescribed tolerances, stop. Otherwise go to step 2.

EXPERIMENTAL VERIFICATION

One of the first core heat exchanger types simulated using the model described above was the warm core of the main heat exchangers in an air separation plant. Table 1 describes the core's physical characteristics. The thermal profiles predicted by the model are shown in Figures 4 and 5. Figure 4 shows the bulk fluid temperature for the outer air core passages versus distance along the core's length. Note the large temperature gradient that develops at the core's midpoint. This effect is more apparent in Figure 5 which shows the air passage fluid temperatures at the core's midpoint versus passage number. The unexpected large gradient predicted at this position as well as the curve's unusual shape led us to develop an experimental program aimed at validating the model.

In an operating plant, 13 thermocouples were inserted through the sidebars of a core identical to the one discussed above. These thermocouples were located at approximately the core's midpoint. The measured temperatures are shown in Figure 5 along with the model predicted temperatures. The measured profiles are not exactly as predicted but the gradient and the shape of the curve are quite similar to the model predictions. Additionally, since the heat exchanger was in cryogenic service and was exposed to the atmosphere in order to insert the thermocouples, a layer of frost formed at elevations where the sidebars were approximately 30 F. Figure 6 is a photograph of the sidebar face. The frost line represents an isotherm across the core face whereas the previously referenced plots show temperatures at a plane of constant distance. The similarity in the isotherm's gradient and shape strengthens the model's credibility. Furthermore, subsequently measuring other operating cores with different passage arrangements has additionally confirmed the model's validity.

CONCLUSIONS AND DISCUSSION

Based on the experimental measurements described above and on using the model to simulate all Air Products core heat exchanger designs, we have concluded that the model adequately represents exchanger performance and that the successive approximation method very reliably solves the model equations. Furthermore, passage arrangement controls overall heat transfer performance and internal temperature profiles.

In addition to representing passage arrangement and specified flow distribution, the model and solution procedure have been extended to handle blocked or partially blocked passages, fouling coefficients, heat leak from the cap sheets, longitudinal conduction in the parting sheets and an approximation for cross-flow arrangements. Distributor and pressure drop calculations based on the converged temperature profiles are also included in the simulation module. The result of our work is a highly versatile and reliable tool which is used to simulate and analyze all Air Products designed cores prior to final design. The simulation is used to verify the required heat transfer length and to generate internal (passage) temperature profiles which are used to analyze thermal stresses that may be induced in the core. The latter effect, particularly important in reversing exchanger service where the core is constantly moving between steady state conditions, can cause premature core rupture for badly prescribed passage arrangements. This was actually observed in several operating cores. We feel that simulation analysis using the tool described here results in better designed exchangers with respect to both heat transfer performance and mechanical integrity.

Table 1. Warm Core Characteristics

Fluid:	A,B,C	D
Fin type:	1/8" lanced	11.5% Perf
Fin spacing:	18 fins/in.	15 fins/in.
Fin thickness:	0.008 in.	0.008 in.
Fin height:	.375 in.	0.375 in.

Core length: 9.33 ft.

Passage arrangement

AB (ABC) ABD AB ABD (ABC) AB AB

A:Air B:Waste C:Product D:Instrument Air

NOMENCLATURES

d	fins per unit length, (in)
f	fin height, in
H	specific enthalpy, BTU/mol
h	heat transfer coefficient, BTU/hr-ft -F
i	subscript denoting parting sheet or passage number
j	iteration counter for successive approximation
k	thermal conductivity, BTU/hr-ft-F
m	molar flow rate, mol/hr
n	number of passages in exchanger
p	damping factor defined in equation (5)
T	fluid temperature, F
TP	parting sheet temperature, F
t	fin thickness, in
W	core width, ft
z	distance in flow direction, ft

CITED LITERATURE

Braun, M., Differential Equations and Their Applications, 2nd ed., pgs 64-76, Springer-Verlag, New York (1975).

Conte, S.D. and de Boor, C., Elementary Numerical Analysis, 3rd ed., pgs. 294-310, McGraw-Hill, New York (1980).

Kao, S., "A Systematic Design Approach for a Multistream Exchanger With Inter-connected Wall", ASME Paper 61-WA-255, 1961.

Kays, W.M. and London, A.L., Compact Heat Exchangers, The National Press, Palo Alto, California (1955).

Perry, R.H., Chilton, C.H. and Kirk-patrick, S.D. eds., Chemical Engineers' Handbook, 4th ed., Chap. 2 pgs 59-61, McGraw-Hill, New York, (1963).

Weimer, R.F. and Hartzog, D.G., "Effects of Maldistribution on the Performance of Multi-Stream, Multi-Passage Exchangers", paper presented at Thirteenth National Heat Transfer Conference, Denver, Colorado, Aug. 6-9, 1972.

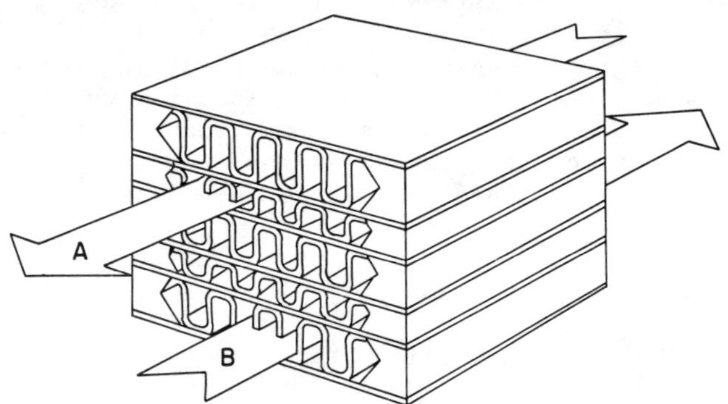

Figure 1. Core heat exchanger with parallel flow.

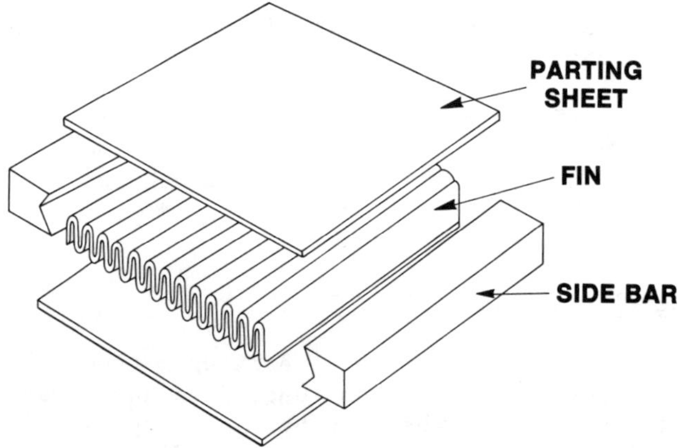

Figure 2. Core heat exchanger passage.

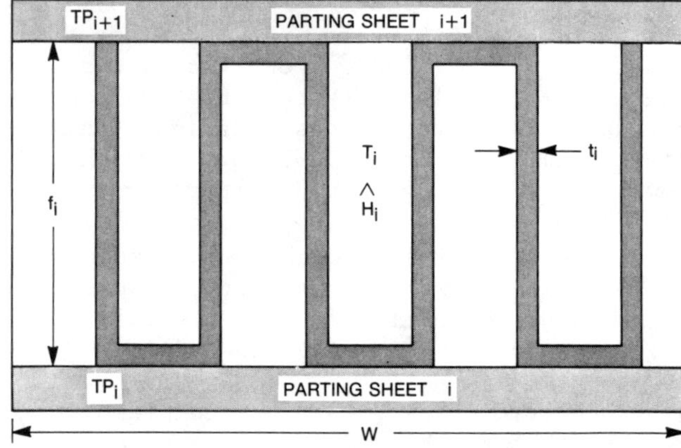

Figure 3. Idealized passage cross section.

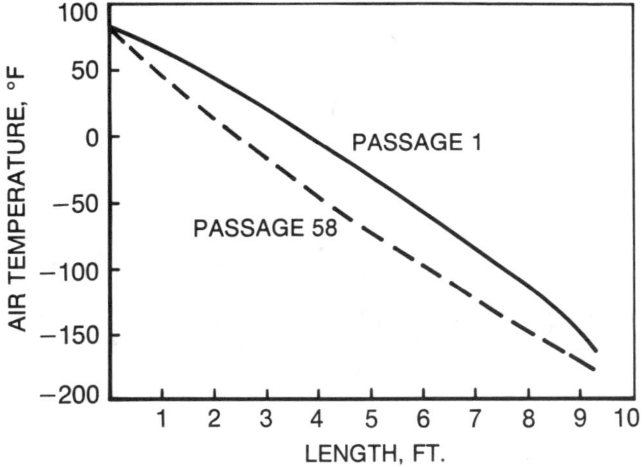

Figure 4. Outer air passage temperature profiles.

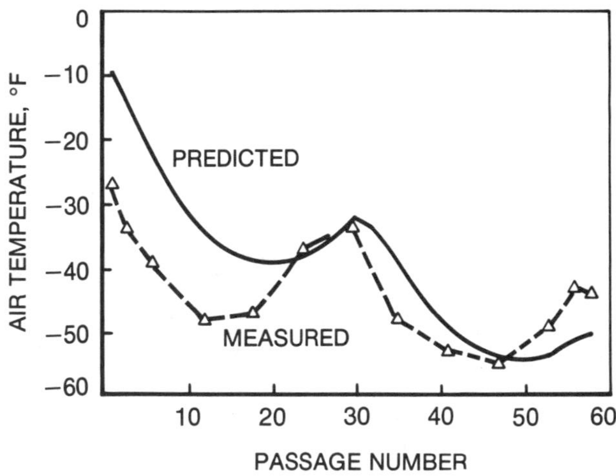

Figure 5. Air passage temperature profiles at exchanger midpoint.

Figure 6. Isotherm across core face.

ANALYSIS AND CORRECTION OF DESIGN DEFICIENCIES IN A 500 T/D OXYGEN PLANT

J. T. BERNSTEIN

Cryogenic Consulting Service, Inc.
Westport, Connecticut

PLANT HISTORY

An air separation plant designed for production of 500 tons per day of gaseous oxygen as well as nitrogen, argon and liquid products was constructed in 1969 for the Burdett Oxygen Company; and was later purchased and is currently owned and operated by Northern Petrochemical Company. Cryogenic Consulting Service was retained initially by Burdett and later by Northern Petrochemical Company to provide consulting as well as engineering and design regarding the analysis and correction of deficiencies in this plant.

The plant consisted of a typical low-pressure reversing heat exchanger air separation unit designed to produce the gaseous oxygen, nitrogen and argon, plus a 200 T/D recycle nitrogen liquefier. Figure 1 is a block flow diagram which shows the overall relationship of the plant components.

Figure 2 is a simplified representation of the basic air separation unit low pressure reversing exchanger process utilized to separate air into the pure gaseous products.

Initial Symptoms and Problems

Some of the initial problems and symptoms exhibited by the plant after its startup included:

- Liquid production was substantially less than design; approximately 40-50 tons below the nominal 200 T/D expected.

- The reversing heat exchanger in the air separation unit exhibited an unusually large warm-end temperature difference; the air to waste nitrogen temperature difference was about 10°-12°F; nitrogen product to air temperature difference about 15°F and oxygen to air temperature difference about 30°-35°F, for an overall average warm-end temperature difference of about 15-16°F. (This is in contrast to a "normal" warm-end temperature difference in a typical reversing heat exchanger plant of about 6-7°F.)

After some period of operation, it was observed that in operation of the basic air separation unit without the liquefier, the air separation plant actually consumed 30-35 T/D of liquid oxygen and was incapable of "stand alone" operation. There was a continual circulation of liquid oxygen from the air separation unit tower to the liquid oxygen storage vessel and back through a hydrocarbon adsorber, and this circulation provided a "connection" initially overlooked but later clearly observed to allow liquid oxygen to be supplied to the air separation unit. Refrigeration in the form of liquid nitrogen is supplied from the recyle liquefier to the air separation columns, and the net excess of liquid available in the columns is produced to storage as either liquid oxygen or liquid nitrogen. Thus, the flux of liquid products between the air separation unit distillation towers and the liquid product storage tanks initially obscured the air separation unit refrigeration loss, and the deficiency in liquid production was initially thought to be a problem of the liquefier unit.

Symptoms and Possible Causes

After a period of initial sorting out of operating experience the symptoms of excess refrigeration loss in the air separation unit were:

1. The unusually large temperature difference between reversing heat exchangers, averaging about 15°F, or more than twice the normal or expected warm-end heat exchanger temperature difference.
2. Consumption of approximately 30 T/D liquid oxygen for "makeup" refrigeration.

These two symptoms were consistent: The excess refrigeration loss at the warm-end of the reversing heat exchangers, about 8.5° higher than normal was equivalent to a refrigeration loss of about 60 btu/mole of feed air. The 30 T/D of liquid oxygen supplied was also equivalent to an excess refrigeration supply of about 60 btu/mole air feed.

Figure 3 shows the normal approximate refrigeration balance for a low pressure cylce oxygen plant, and the excess refrigeration loss manifested in the heat exchangers and supplied by the liquid oxygen. The normal refrigeration balance indicates that the low pressure oxygen cycle is supplied with feed air approximately 6 ATMA, and produces gaseous oxygen, gaseous nitrogen and "waste" nitrogen at 1 ATMA. Refrigeration is required to sustain this process because of the temperature difference between the air feed and low pressure gas products (normally about 6°-7°F), plus the heat inleakage from ambient temperature to the low temperature process equipment through the cold box insulation. Normally this refrigeration requirement is made up by expanding approximately 10-13% of the air feed through a low temperature expander from a pressure of 6 ATMA to 1.3 ATMA.

The normal low pressure air separation process is therefore self-sufficient in terms of refrigeration, and under condition of increased expander flow can actually produce small quantities of liquid oxygen or nitrogen product. In the case of the 500 T/D plant in question, no increase in expander flow within the process and equipment operational parameters provided

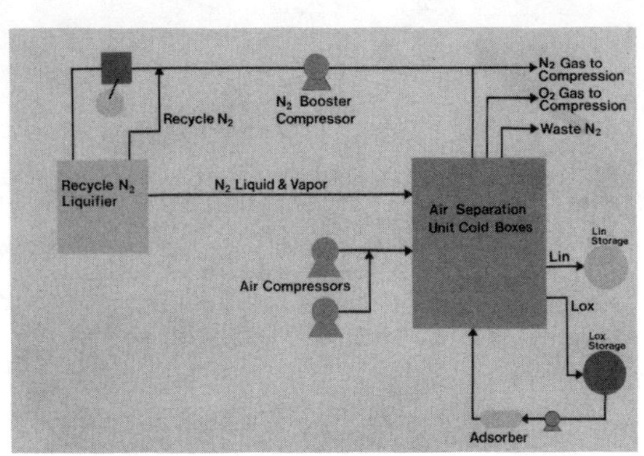

Figure 1. Morris 500 T/D air separation unit & liquefier.

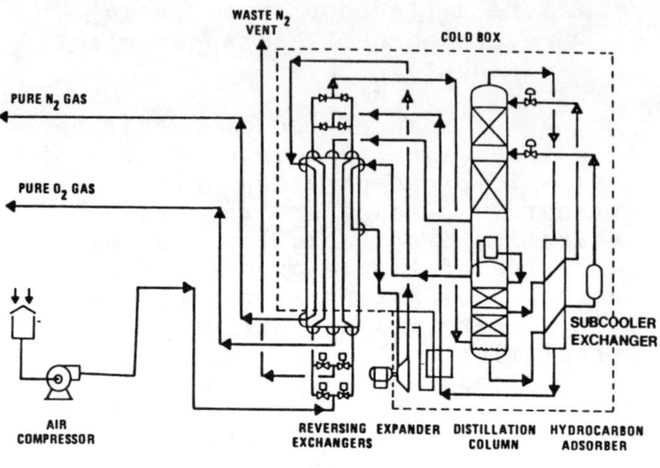

Figure 2. Low pressure cycle oxygen plant with reversing exchangers.

	"Normal" 500 T/D O_2 btu/mol air feed	Morris 500 T/D O_2 Plant btu/mol air feed
REFRIGERATION REQUIRED:		
1) Heat Exchanger Warm End Enthalpy Difference	40	100
2) Heat In-leak	15	18
	55	118
REFRIGERATION SUPPLY:		
1) L.P. Expander	55	60
2) Liquid Oxygen	0	58
	55	118

Figure 3. Approximate heat balance -

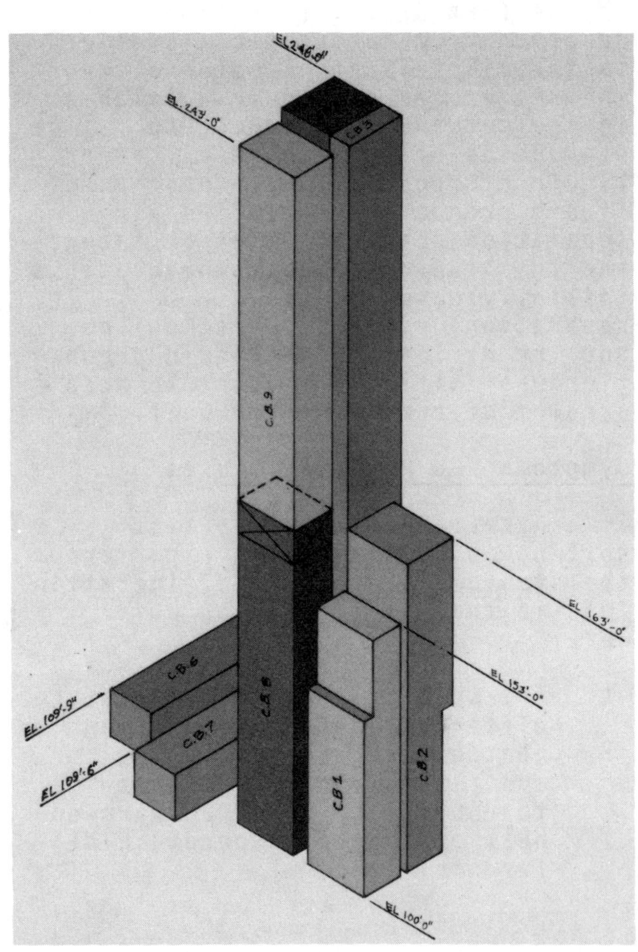

Figure 4. 500 T/D oxygen plant cold box arrangement.

sufficient refrigeration for normal plant operation without supplemental liquid oxygen supply. The plant was provided with duplicate low pressure expanders, the second unit to be used for start-up or as a spare, but operation of both expanders would not noticeably affect the excess refrigeration loss. The observed refrigeration requirement was in fact nearly double that of a "normal" plant.

Possible Causes of Refrigeration Loss

The possible causes of refrigeraton loss intially considered by the plant operators were:

1. Reversing Heat Exchanger Performance:

   The unusually large warm-end temperature difference (more than twice normal) appeared to indicate poor performance by the reversing heat exchangers.

2. High Heat Inleakage:

   The plant construction utilized a large number of individual "cold box" sections designed to be transporatable by truck shipment. This configuration resulted in a larger than normal cold box area and a larger than normal heat leak. In addition, the high purity nitrogen and argon towers were situated so as to require cryogenic liquid pumping rather than the "normal" gravity flow transfer of liquid between towers.

Analysis of Possible Causes

In our analysis, excess plant heat leak was ruled out as a major cause of the 30 T/D liquid oxygen eqivalent refrigeration loss. As indicated in Figure 3 the normal expected heat leak for a plant of this size and configuration is about 15 btu/mole of air feed. The particular design and configuration involving multiple cold boxes and liquid pumps by analysis could have resulted in an excess heat leak in the range of 10-20%, but in no way could have accounted for the excess loss equivalent to four times the expected heat leak. Other possibilities for refrigeration loss included physical leaks, particularly leaks of liquid oxygen or nitrogen. Although the cold box showed some frost spots on the external structure, no gross physical leaks were evident which could have accounted for the large excess refrigeration loss.

Reversing Heat Exchanger Performance

The reversing heat exchangers were typical air separation unit made exchangers as used during the period of plant construction. Sixteen parallel brazed aluminum plate-fin type exchangers were used, each approximately 2 ft. X 2 ft. in cross section, with two exchangers in series each, about 8.5 ft. in length. The heat exchangers had a peculiar construction in that the pure product oxygen and nitrogen were withdrawn in "stacked" side headers (see figure 5).

Analysis of the heat exchanger design indicated that some degree of difference between oxygen, nitrogen and waste product temperatures was expected for the design conditions, but that the heat exchanger design and surface provided should have been adequate. However, this analysis is based the normal design flows in the exchanger in which the total product flow of oxygen, nitrogen and waste is essentially equal to the quantity of air feed in the exchanger. In actuality, since 30 T/D of liquid oxygen was added to the plant, the product streams exceeded the air feed by about 1%.

The mass unbalance created by the excess 1% of product stream flow changes the temperature duty curves for the reversing heat exchangers (see figure 6) causing a larger than normal temperature difference at the warm-end of the exchangers. In other words, given the supply of 30 T/D of liquid oxygen as refrigeration make up to the plant, the reversing heat exchangers would have to exhibit a larger than normal warm-end temperature difference even though they were perfectly adequate with respect to their design, heat exchange coefficient and surface area. Thus, it was not clear whether the unusually large temperature difference at the heat exchanger warm-end was a cause or effect of the excess refrigeration loss. Assuming that the heat exchangers supplied were in accordance with the actual design and were not excessively fouled with CO_2 or water ice, calculations indicated that the heat exchanger performance should have been adequate, and that heat exchanger design was not probable cause of the excess refrigeration loss.

The analysis of the actual heat exchanger performance was hampered by missing or inaccurate thermocouples, particularly at the cold end of the exchangers, and the problems of analyzing the conditions for 16 separate heat exchanger units operating in parallel. As a result, the confirmation of adequate heat exchanger surface was gained largely by calculation, not by actual performance.

Subcooler Performance

Observation of the conditions around the reversing heat exchangers indicated another unusual condition: the product streams leaving the subcooler exchangers and entering the cold end of the reversing exchangers were at a temperature of about -288°F, as opposed to a normal or typical temperature of -280°F. The air separation products leaving the columns, consisting of oxygen, waste nitrogen, and pure nitrogen are normally warmed up in heat exchange with liquid nitrogen reflux and enriched air leaving the "high pressure" tower. The one reason for warming up the products is that product stream temperatures below -280°F entering the reversing heat exchangers may cause the air leaving the reversing heat exchangers to liquefy. The construction of the reversing exchangers is such that the liquefied air may physically be retained in the exchangers and upon reversal of the exchangers be transferred into the waste gas stream, resulting in a loss of refrigeration as well as feed air.

Another reason for warming the product streams in the subcooler exchangers is that this warming is required to bring about a refrigeration balance for the tower system. Because it is undesirable to form liquid air within the reversing heat exchangers, the air temperature exiting the reversing heat exchangers is normally limited to being not colder than 2° above the dew point of the air leaving the reversing heat exchangers (or about -275°F).

Figure 7, a Mollier chart for air, shows that in order to have the same enthalpy as air entering the column system at about 90 psia and just about the dew point, "air" or product gases leaving the tower system at about 18 psia would have to warmed to a temperature of about 35° above the product dew point. Thus the "subcooler exchangers" are absolutely required in the low pressure reversing exchanger plant in order to bring about an enthalpy balance between entering air feed and leaving product streams. Under normal conditions, the enriched air liquid stream leaving the bottom of the high pressure tower is at it's

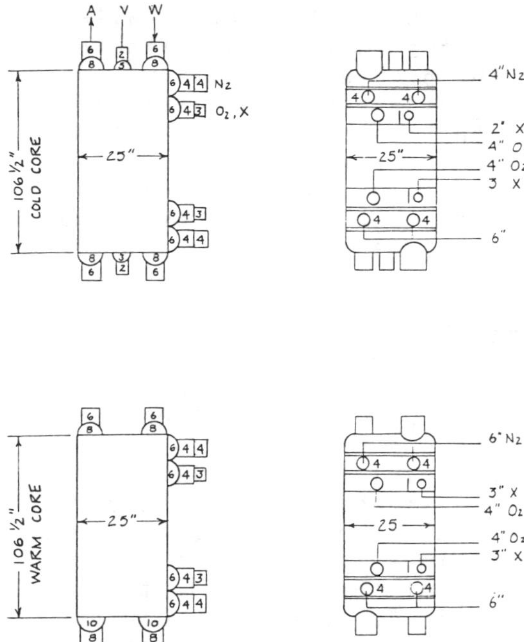

Figure 5. Original reversing heat exchanger configuration.

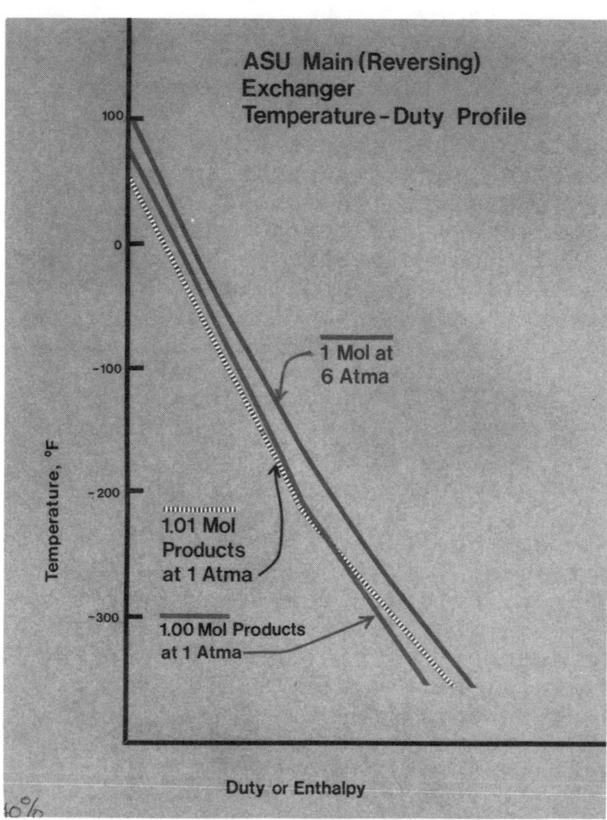

Figure 6. Effect of mass unbalance on reversing exchanger temperatures.

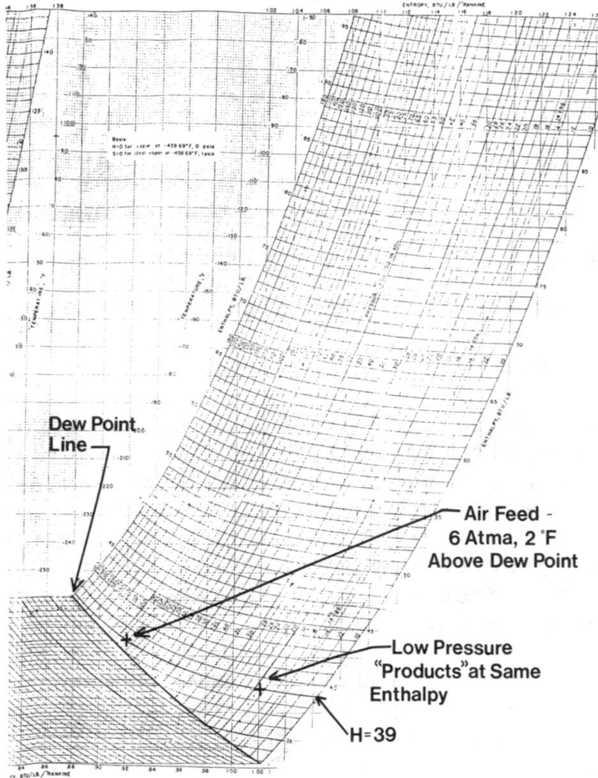

Figure 7. T-S diagram showing need to superheat column products.

Figure 8. Cold box during erection. Subcooler exchangers are at top.

saturated liquid temperature of approximately -276°F and a 4° approach to this temperature warms the product streams up to -280°F. In the 500 T/D oxygen plant in question, the performance of enriched air section of the subcooler exchanger was obscured by the expander inlet heat exchange section of the subcooler. Once the expander inlet precooling section was temporarily taken out of service, the available temperature indicators at the warm-end of the subcooler indicated product stream temperatures of about -288°F, or significantly colder than the "normal" -280°F.

Heat balances around the tower section of the air separation unit clearly indicated that temperature of -288°F for the products leaving the subcooler exchangers represented a sufficiently low enthalpy to require addition of 30 T/D of liquid oxygen to maintain the column liquid levels and refrigeration balance. The addition of the liquid oxygen effectively vaporized in the column system to maintain refrigeration balance provided a mass imbalance in the reversing heat exchangers, resulting in the unusually large warm-end temperature difference.

The search for the reason for poor subcooler performance yielded similar results to the analysis of the main exchangers. Based on the subcooler configuration, surface provided and calculated coeffiecients, the subcooler should have provided adequate performance. As with the reversing exchangers, the possibility of CO2 and water ice fouling was considered, and the plant was thoroughly defrosted on a number of occasions without noticeable improvement in performance. The subcoolers were, like the main exchangers, composed of a number of parallel brazed aluminum exchangers units and different product streams were allocated to different exchanger cores with four different parallel paths for the enriched air liquid flow. The problem of optimum balancing of the rich air liquid flow to the subcoolers was considered on a theoretical and operational basis without any significant result in improved subcooler performance.

The cause of poor subcooler performance was finally discovered in a somewhat unexpected area. After analysis of temperatures, flows, and operating additions and calculation of exchanger design parameters yielded no positive results, it was observed that the physical location of the subcoolers was approximately 45 feet above the base level of the air separation unit. (This "observation" could only be made by looking at the plant construction drawings, as all of the heat exchangers, towers and piping are enclosed in the insulated "cold box".) The enriched liquid air, flowing from the bottom of the lower tower, at approximately the cold box base level, where the saturated liquid temperature is -276°F at a pressure of about 85 psia would be reduced in pressure by about 16 psia by the change in elevation alone. The combination of elevation change and pressure drop was estimated to amount to 20 psia, and the lowering of the enriched liquid air pressure by 20 psia will result in a lowering of saturated liquid temperature to about -284°F. The expected 4° approach to the -284°F enriched liquid air temperature at the lowered pressure, together with the expected 4° warm-end approach for the subcoolers is consistent with the observed temperature of -288°F for the products leaving the subcooler exchanger.

Figure 8 shows the general configuration of the reversing heat exchangers, mounted in the bottom of the heat exchanger cold box, with the subcoolers mounted above. The size of the reversing heat exchanger bank was such that it could be mounted into a truck-shippable 14 ft x 9 ft cross section and it is speculated that in order to maintain this shippable dimension, design engineers not familiar with the process aspects and implications of heat exchanger placement located the subcoolers above the reversing heat exchangers. This example shows that a lack of communication between process designers and mechanical designers of a cold box unit can result in serious problems and deficiencies.

A project is currently nearing completion which will replace both the reversing heat exchangers (which have developed leaks) as well as the subcoolers. In the new installation, the subcoolers are being located close to the level of the bottom of the columns. In order to minimize plant down-time, a new exchanger cold box is being constructed along side the existing unit.

ADVANCED HEAT TRANSFER TECHNOLOGY IN CRACKED GAS SEPARATION

P. S. O'NEILL
G. W. FENNER

Union Carbide Corporation
P.O. Box 44
Tonawanda, New York

E. F. WILLIAMSON

Union Carbide Corporation
P.O. Box 50
Hahnville, Louisiana

Thermal design and operating aspects are discussed for an integrated cold box and shell-tube heat exchangers in the demethanizer chilling train of a large ethylene plant. Factors such as component integration, flow splitting, and two-phase entry are reviewed, along with operating comments.

Heat exchange, distillation, and heat recovery at low temperatures are essential operations in modern ethylene and light hydrocarbon processing plants. In an olefins plant, for example, feed streams consisting of a pyrolyzed naphtha, LPG or ethane are chilled using various levels of propane or propylene and ethylene refrigeration from ambient temperature to about -100°C. Recovered condensates, which are rich in the desired ethylene product, are further fractionated. Hydrogen and methane streams are further processed in a "cold box" to recover the hydrogen at temperatures from -100°C to about -160°C.

Heat exchangers in these services must be able to handle large flows with close temperature approaches and high thermal loads per unit of heat exchanger volume. Streams are often two-phase, and compact arrangement and packaging of components is essential to minimize equipment costs and thermal losses.

A "cold box" generally consists of a cylindrical steel casing, having a minimum number of seals and penetrations, and housing a number of cryogenic (plate-fin) heat exchangers, phase separators, small distillation columns and piping. The unit is filled with powdered insulation and dry inert gas to minimize heat losses. Piping flexibility required during cooldown to cryogenic temperatures is provided wholly within the casing structure, thus minimizing field piping costs. Heat exchangers in an ethylene plant demethanizer chilling train perform four functions:

Recovery of ethylene rich liquids from the feed

Upgrading and recovery of hydrogen product

Rejection of methane-rich fuel streams

Recovery of refrigeration at various temperature levels

This paper discusses a specific example of the use of a cold box and High Flux heat exchanger tubing, both designed and manufactured by Union Carbide Corporation, in the demethanizer chilling train of the Lummus designed Union Carbide olefins plant at Taft, Louisiana. Although both High Flux tubing and cold boxes have been extensively used in other olefins plants (Ref. 1,4), the Taft plant is the best example of extensive integration of both enhanced surface shell-tube, and plate-fin heat exchangers in a demethanizer chilling train. A key advantage of the High Flux exchangers is the small bundle diameters made possible by overall heat transfer coefficients approximately 2-3 times greater than corresponding bare tube surface. The small High Flux exchanger size made it possible to reliably handle all of the partially condensed streams from the cold box without resorting to expensive two-phase entry or distributor devices at the exchanger inlet, thus minimizing vessels. The design of the plate-fin exchangers within the cold box (some of which were in parallel) could therefore be simplified, due to the absence of two-phase feeds above temperature levels of -100°C.

The equipment discussed in this paper has been in operation for about 4 years, and has at times operated above design capacity. Subsequent sections of the paper will discuss the layout and arrangement of the exchangers and cold box, design aspects, and operation of the integrated system.

DISCUSSION

The plant cracks light naphtha feedstocks which typically produce a broad spectrum of olefinic, aliphatic, and aromatic hydrocarbons, in addition to hydrogen. Recovery of the desired ethylene product, which is present in the feed at a concentration of 25-30 mol %, involves a series of systematic distillation steps to recover hydrogen, methane, ethane, ethylene, propylene, and heavier hydrocarbons. Heavy aromatics and fuel fractions (pyrolysis gasolines) are first removed from the cracked gas followed by drying and compression to about 35 atm pressure. This paper will focus on that section of the plant where the dry cracked gas is demethanized, and the recovered hydrogen streams are upgraded to a higher purity product.

(A) Flow Arrangement

A simplified flow sheet of the demethanizer chilling train is shown in Figure 1. The system consists of 6 High Flux shell-tube, U-bundle, exchangers (Items E-1 to E-6) which vaporize ethane, propylene or ethylene, and partially condense a gas stream consisting of hydrogen, methane, C_2, C_3 and C_4 hydrocarbons. The shell-tube exchangers receive partially condensed streams from the cold box, where multi-stream plate-fin exchangers recover heat from cold return and recycle streams originating at the demethanizer. The water free plant feed is first cooled to about -10°C from heat exchange at the ethylene fractionator column, where about 20% of the feed condenses. The two-phase stream is then fed to the High Flux ethane vaporizer E-1, which converts the liquid ethane product to vapor prior to recycle to the cracking furnace. After heat exchange at E-1, the feed contains about 35% condensate, and enters exchanger

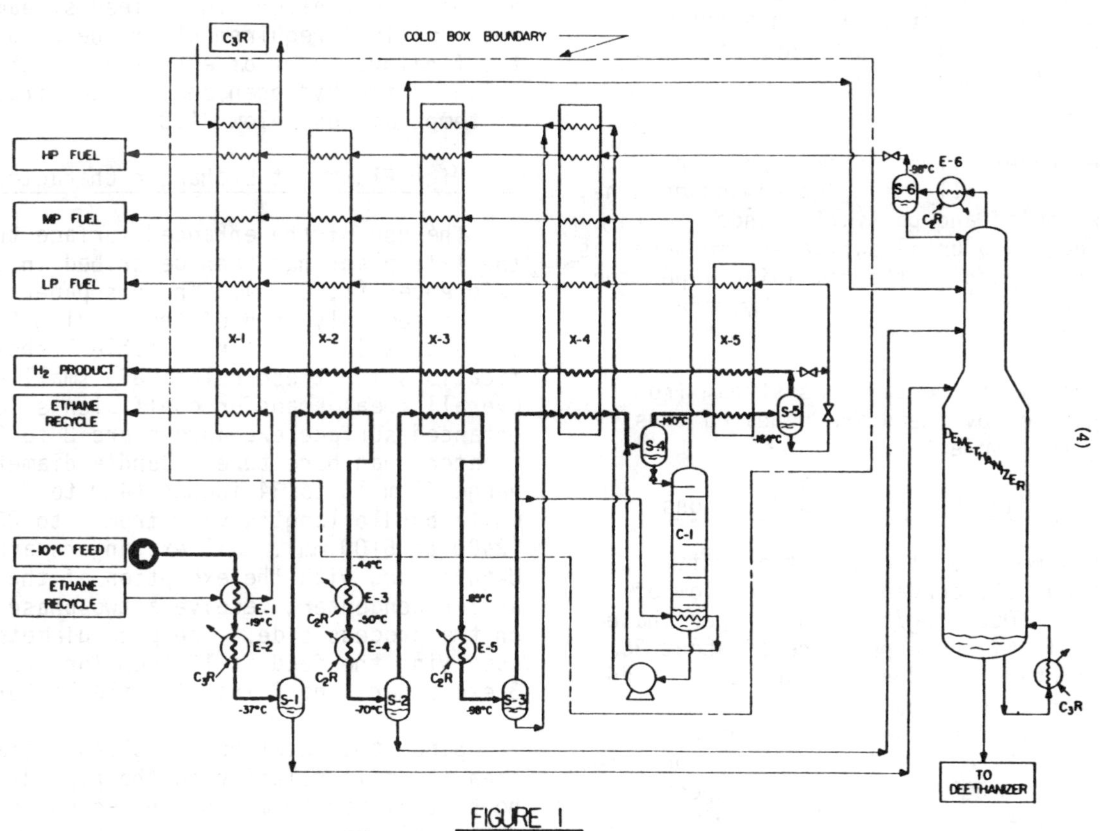

FIGURE I
DEMETHANIZER FEED CHILLING TRAIN
UNION CARBIDE CORP., TAFT, LA.

E-2, where further condensation occurs, using propylene refrigeration at -40°C. At the outlet of E-2, the condensed fraction is removed and sent to the demethanizer column as a feed stream, while the vapor fraction is sent to the cold box plate-fin exchanger X-2. In X-2, about 10% of the charge gas condenses in heat exchange against the return hydrogen and fuel gas streams. The two-phase stream then leaves the cold box and enters High Flux exchangers E-3 and E-4, where 2 stages of ethylene refrigeration at -50°C and -75°C chill and further condense about 30% of the remainder of the feed. At the exit of E-4, the condensate is again separated and the vapor portion returned to the cold box. At the second separator, almost 75% of the original feed stream has been condensed. The cold box exchanger X-3 condenses about 25% of the remaining vapor portion against the hydrogen and fuel gas streams, and the partly condensed stream is sent to the final High Flux chilling train exchanger E-5, where ethylene refrigeration at -100°C is used. Again, the vapor protion is returned to the cold box, chilled and partially condensed against the hydrogen and fuel streams to about -140°C. The ethylene-rich liquid phase is processed, after throttling, in a stripper column, which produces a relatively high purity (∿80%) ethylene product at a high ethylene recovery. This stream is combined with the liquid phase from S-3 to become the final demethanizer feed.

Vapor phase effluent from separator S-4, consisting essentially of hydrogen and methane, is further cooled and partially condensed in heat exchanger X-5 until sufficient methane has been condensed to give the remaining vapor the desired hydrogen concentration. Hydrogen product is recovered from the vapor phase at S-5. Finally, the hydrogen product and fuel streams along with the tail gas stream from the demethanizer overhead are reheated thus completing the cycle.

(B) Heat and Mass Balance Considerations

The cracked gas stream at the system inlet (E-1) is about 20% condensed, has a flow of approximately 100000 kg/hr with a vapor phase mol weight of 26, and a pressure of about 35 bar. Considerable condensate can be produced from the feed at various cooling levels. At the -40°C level (first separator) for example, approximately 60% of the feed will be condensed. At the -70 and -100°C levels, the respective amounts of condensate are at 80 and 90%. The largest fraction of the refrigeration for the system is used in the High Flux shell-tube exchangers rather than inside the cold box. This is a consequence of the fact that the heat contents of all the return streams in the cold box are small relative to that of the charge gas feed. At the -100°C level, for example, only about 12% of the total generated condensate is produced in the cold box exchangers. Distribution of the refrigeration sources for the condensate produced from the charge gas feed is as follows:

Source	% of Total
Cold box return streams	12
Upstream of system inlet	22
Ethane recycle vaporization	14
C_3H_6 refrigeration	28
Ethylene refrigeration	24

In the above table, the first 3 refrigeration sources are in effect "free" while the last 2 refrigeration sources are produced by a multi-stage system which consumes energy. It is apparent from the table, that although the cold box is a large vessel containing several multi-stream heat exchangers, with extensive exchange among the various streams, it still provides only a minor amount of the total refrigeration requirement for production of the demethanizer liquid feed streams. The most critical requirement of the cold box is the final recovery of ethylene and the separation of the hydrogen and methane fractions at temperatures below -100°C.

(C) High Flux Heat Exchanger Characteristics

The use of the enhanced surface tubing in the Taft plant has been described in previous references (1, 2, 3). In this paper attention is focused only on 6 of the 20 High Flux exchangers in the plant. Table 1 shows specifications for these relatively small exchangers. Overall, heat transfer coefficients for the enhanced surface exchangers are 2 to 3 times greater than bare tube. Bundle diameters range from 16 to 34 inches (400 to 850 mm), while bundle lengths vary from 8 to 20 ft (2400 to 6100 mm). All exchangers are 2 pass U-tube, and with the exception of the demethanizer condenser, receive a two-phase stream on the process side. The pipe diameter supplying the feed is 10 inch for the first 3 exchangers and 8 inch for the latter.

Since the inlet heat exchanger channel area is small relative to the pipe diameter, no special requirements are needed to assure proper distribution of liquid and vapor to the tubeside. Velocities at all points in the inlet region are high enough to maintain

TABLE 1

HIGH FLUX HEAT EXCHANGER SPECIFICATIONS

ITEM NO.	HEAT DUTY MW	SURFACE AREA M^2	EXCHANGER SIZE (CM x CM x CM)	COLD END ΔT ^oC DESIGN
E-1	1.5	92.44	71/132 x 244	7.8
E-2	3.4	367.9	86/175 x 610	2.8
E-3	0.67	92.9	58/127 x 366	2.2
E-4	1.9	162.6	64/127 x 549	2.8
E-5	0.7	69.2	41/102 x 610	2.8
E-6	1.26	205.3	64/122 x 701	2.8

homogeneous dispersed flow. Separation or mal-distribution of condensate is more likely when the inlet nozzle is small relative to the flow area in the first pass. If mal-distribution occurs, the vapor phase condenses separately, without the original liquid phase, and results in less condensate production for a given outlet temperature. Phase separation can have serious consequences in condensation of multi-component mixtures in heat exchangers designed for low outlet approach temperature differences. It was not necessary to use shell-tube flow devices, or reverse-feed the shell-tube exchangers to insure proper distribution of liquid and vapor. The ability to handle two-phase flows without liquid mal-distribution resulted in the maximum liquid recovery for the demethanizer feeds, and a minimum of phase separators and associated piping. A second advantage of using enhanced surface shell-tube exchangers was simplication of plant layout. None of the High Flux services in the chilling train, or in fact the entire plant, required parallel units. Single units in series reduces instrumentation, controls and piping costs relative to parallel units. The Taft plant has an ethylene capacity of approximately 350,000 metric tons/year.

(D) Cold Box Characteristics

The overall specifications of the cold box for this application is presented in Table 2. Overhead vapor at -98°C from separator S-3, which still contains appreciable quantities of ethylene, is further processed in the cold box to recover this product. This stream is cooled to -140°C by heat exchange against the warming return streams in heat exchanger X-4. The resultant liquid phase, which contains the bulk of the ethylene, is throttled and processed in a secondary demethanizer (stripper column). The purpose of this column is to produce a relatively high purity (~80%) ethylene product while achieving high ethylene recovery and thus unload the primary demethanizer column. Ethylene product purity is controlled and limited by the temperature level to which the column bottoms can be reboiled. Ethylene recovery is determined by the operating temperatures of separator S-3, which defines the amount of ethylene lost in S-3 overheads. Liquid ethylene product is pumped and combined with the liquid phase from S-3 prior to further processing in the demethanizer.

TABLE 2

COLD BOX SPECIFICATIONS

OVERALL SIZE - 30.9 M HIGH
3.7 M DIAMETER UP TO 20.6 M AND
2.4 M DIAMETER THEREAFTER
NUMBER OF PLATE-FIN EXCHANGERS - 5
NUMBER OF PHASE SEPARATORS - 5
COLUMN SIZE - 99 CM DIAMETER

Hydrogen recovery is a critical function of the cold box. The overhead vapor of S-4 is further cooled and partially condensed in heat exchanger X-5 to a temperature level consistent with the desired hydrogen purity. The methane-rich liquid stream from S-5 is throttled to a pressure which will allow the methane (fuel) to boil at the temperature needed to provide the required driving force (ΔT) to cool the feed to the S-5 temperature.

Raising the fuel pressure has a direct impact on the purity of the hydrogen product because the liquid methane will boil at a higher temperature, which will cause S-5 temperature to increase, thus lowering the hydrogen product purity. A higher hydrogen product purity can be achieved by reducing the boiling temperature of the liquid methane. This can be accomplished by either lowering the fuel pressure or reducing the partial pressure of the methane in the fule stream by mixing a portion of the hydrogen product. Thus, there is a trade-off between the hydrogen product purity, hydrogen recovery, and fuel pressure for any given fixed feed composition and feed pressure. Achievement of the low temperature necessary for obtaining 95% hydrogen is insured by maintaining low heat leak into the low temperature portion of the cold box and by careful design of a two-phase entry device for proper distribution of the liquid and vapor fractions of the fuel stream into heat exchanger X-5. Uniform distribution is accomplished by phase separating the fuel stream and recombining the stream internally in the heat exchanger to insure uniform distribution from passage to passage.

(E) Design and Optimization Considerations

The process previously described is one of many alternative schemes for separation of methane and hydrogen from cracked gas. General analysis of demethanizer chilling trains, identification of key variables, and discussion of the various cycles is given in Reference 6. The Taft demethanizer column can be classified as one of high pressure. Intermediate and low pressure (10 bar) column pressures are used in some olefins plants. These conserve ethylene refrigeration, but result in use of more methane refrigeration below -100°C. The key advantage of the cycle described here is exceptionally low (0.2%) ethylene product losses, while minimizing the use of refrigeration below -100°C, and providing maximum operating flexibility in adjusting to feed stock changes. The use of the internal column for ethylene recovery, and operating at temperature approaches of only 2-3°C on the High Flux demethanizer condenser, for example, reduced product ethylene composition in the demethanizer waste gas stream to only hundredths of a percent. The overall system was designed to recover about 85% of the hydrogen (as a 95% stream), and 97% of the ethylene. Feed stock variations were allowed for in the design, with hydrogen ranging from 27-39%, methane from 26-41% and ethane from 3 to 8%, at the cold box entrance. Actual ethylene recovery exceeded 99%.

Some further improvements could have been made in the system process/equipment optimization to reduce sizes of the more expensive components. For example, if High Flux exchangers E-2, 3 and 4 were doubled in size the cold end temperature approaches could be reduced about 1.5°C. In general, reduction of temperature differences on the shell-tube exchangers will help to either unload the ethylene refrigeration system or improve recovery of ethylene product. It is believed that the optimum cold end approach temperature difference would be about 1.4°C for the shell tube (High Flux) exchangers in the chilling train, as opposed to actual design values of about 2.8°C.

(F) Operating Comments

Following startup in late 1978, the plant was subsequently operated at about 110% of design capacity with no significant equipment problems. Some observations made by Operations personnel were as follows:

(1) Temperature measurements indicated that cold box exchanger X-4 performed at higher than design thermal efficiency, helping to reduce ethylene losses in the fuel streams.

(2) Demethanizer condenser E-6 in practice operated at cold end temperature approaches of only 1-2°C. The resulting lower reflux temperature helped to reduce ethylene product losses in the HP fuel stream to essentially zero.

(3) Higher than expected hydrogen recovery resulted from above design performance of exchanger X-5.

(4) Exchanger X-1 experienced some early variations in performance due to water/ice in the propylene refrigerant. The problem was solved by occasional methanol injection during operation.

(5) High Flux exchanger E-5 in early operation experienced higher than expected tubeside pressure drop. The problem was resolved during a scheduled shutdown when excessive debris from plant construction was found in the inlet channel. No further operating problems have been experienced in any of the demethanizer chilling train components.

In conclusion, the equipment described in this paper has operated with a high degree of reliability and thermal efficiency. Ethylene product losses have been negligible and hydrogen recovery excellent in spite of wide variations in plant load and feed stock composition during the past 4 years. The described equipment could be designed and applied with confidence in ethylene plants of much larger size, and with feed stocks derived from LPG, ethane, or naphtha.

LITERATURE CITED

1. O'Neill, P. S., et al, "Application of High Performance Evaporator Tubing in Refrigeration Systems of Large Olefins Plants", AIChE Symp. Series, 76, No. 199, p. 289 (1980).

2. Wett, T., "Flexible Olefins Plant designed for Long Run", Oil & Gas Journal, p. 77, (April 14, 1980).

3. Ibid, 76, p. 83 (Sept. 4, 1978).

4. Davis, J. S., and Martin, J. R., "Cryogenics-The Means for Efficient Synthesis Gas Processing", AIChE 86th National Meeting, Houston, Texas (April 1-5, 1979).

5. Fisher, T. F., "Hydrogen Conservation in Refineries via Cryogenic Processing", NPRA Annaul Meeting, San Antonio, Texas (March 26-28, 1972).

6. Huang, W., and Chou, C. C., "Improve Demethanizer Operation", Hydrocarbon Processing, p. 223 (Sept. 1980).

7. Antonelli, R., and O'Neill, P. S., "Design and Application Considerations for Heat Exchangers with Enhanced Boiling Surfaces", International Conference, Dubrovnik, Yugoslavia (Sept. 1981).

IMPROVED COMPUTER PROGRAM TO PREDICT ROLL-OVER IN LNG WITH AND WITHOUT NITROGEN

A self-contained program requiring no thermo-physical property input has been written to expand the original roll-over model of Chatterjee and Geist. Most significantly, nitrogen, a more volatile component of higher liquid density, is included. The effect of nitrogen on roll-over and the model's validity are discussed in detail.

NIRMAL CHATTERJEE
DAVID W. STUDER
Air Products and Chemicals, Inc.
Allentown, Pennsylvania

Over the last decade, there has been a considerable amount of work on roll-over and its prevention in liquefied natural gas (LNG) storage tanks. Several incidents in the early 1970's started this work; within a year the phenomenon was clearly understood, and operating guidelines that virtually eliminated the problem were established. The main feature was that fill-induced stratification of different density LNG in a storage tank led to roll-over, and therefore avoiding stratification would eliminate the roll-over problem.[1,2,3] The original incidents occured with LNG containing very little nitrogen. At that time, it was pointed out that if the LNG contained a substantial amount of nitrogen, it could spontaneously stratify in an initially homogeneous tank and lead to roll-over.[4] Incidents in a peak shaving plant confirmed this theory and additional work to understand nitrogen induced roll-over was undertaken. The result was a better understanding of all aspects of roll-over, and guidelines for both design and operation of LNG plants were appropriately modified for roll-over free operation.

A mathematical model of roll-over, developed at Air Products in the early 1970's, had been extremely helpful in augmenting plant data and understanding the phenomenon. The model simulated the dynamic behavior of an initially stratified LNG tank (the LNG consisting of two components) and the effect of mixing of the stratified layers on the boil-off rates from these tanks. Once confidence in the accuracy of the model was established, many hypothetical roll-over cases were run on the computer. The results were used in developing the operational guidelines that would eliminate roll-over. When nitrogen induced roll-over was discovered to be a problem, a third component, nitrogen, was incorporated into the program. This considerably increased the size and complexity of the program since accurate calculation of physical properties had to be incorporated for the accuracy of the model, and small concentrations of nitrogen substantially changed the physical properties of LNG.

While roll-over is an operational problem, the program was mainly used in the initial design phase and occasionally if a new source of LNG were considered. In the future, as the sources of LNG multiply and storage tanks have to accommodate LNG's of many different densities, the roll-over program will increasingly be used in the field. Hence the computer program will have to be compatible with small, desk-top or mini-computers and still be accurate. This paper describes how the Air Products' roll-over program was modified to achieve this purpose.

ROLL-OVER: THEORY

Stratification

Roll-over is associated with a sudden unexpected increase in boil-off from an LNG storage tank. This phenomenon occurs when stratified layers of LNG mix. Although LNG's of various composition and densities are miscible, very small density differences between fill and tank liquid can lead to stratification. Measurements have indicated that there is almost a step change in temperature and composition within such a tank (Figure 1). Since the "interface" provides a resistance to heat and mass transfer, all the heat leak into the storage tank does not result in boil-off of liquid. The bottom layer(s) store some of the heat. With time, the bottom layers continue to warm and get lighter. When the densities of two adjacent layers are almost equal, the layers mix relatively quickly, and the superheat stored in the bottom layer is released. This results in a very large increase in boil-off rates or "roll-over".

A tank containing well-mixed and nitrogen free LNG does not spontaneously stratify (Figure 2). The heat-leak induced convection currents in the tank lead to liquid vaporizing at the surface and the resultant heavier LNG sinks to the bottom and a continuous circulation process is established. The liquid, after partial vaporization, is heavier because methane is preferentially vaporized and methane is also the less dense component.

Storage tanks containing an initially well mixed LNG with nitrogen can undergo spontaneous stratification. This is because nitrogen is heavier than methane but boils at a lower temperature than methane. Hence, if one visualizes the normal convection in an LNG tank (Figure 3a), an element of liquid reaching the top of the tank flashes, becomes lighter, and stays at the top of the tank. As this proceeds, the lighter layer on top grows in depth. If the initial nitrogen concentration is high enough (>3%), a point can be reached where the denser fluid no longer has sufficient buoyancy to penetrate this lighter layer, and stratification occurs (Figure 3b). Roll-over is then a matter of time. It is important to recognize that this stratification is spontaneous; nothing can be done during the filling operation to eliminate it.[5]

Interfacial Heat and Mass Transfer

A simple model of roll-over assuming well defined stratified layers with no heat and mass transfer across the interfaces only qualitatively explains the phenomenon. For the model to have any value as an operational guide, the model has to successfully simulate known historical cases of roll-over. We are able to do this by incorporating the heat and mass-transfer that occurs across the interface into the model. The correlations are based on experimental data of Turner.[6] Although Turner's experiments were with salt solutions, it is felt that methane-ethane or nitrogen-methane-ethane mixtures should behave similarly (Figure 4). For both heat and mass transfer, the independent variable is a "stability criteria" representing the ratio of change in density due to temperature and the change in density due to composition. For heat transfer correlation the dependent variable is the ratio of heat flux across the interface to the convective heat flux from a flat plate; for mass transfer the dependent variable is the dimensionless ratio of mass and heat transfer across the interface. The nature of the curves, with their sharp swing up as the stability ratio reaches one, and the sensitivity of the results to small errors in physical properties calculation implies that, for the program to succeed, one has to have very accurate physical and thermal data. For a methane-ethane system very simple correlations are adequate. When nitrogen was incorporated, the initial program used Air Products' extensive computer library of data and correlations. This, however, made the program extremely large and meant that no other computer system could use the program.

ROLL-OVER: COMPUTER MODEL

Program Requirements

Since the program attempts to simulate actual tank performance, the types of storage facilities and their requirements is first examined. There are three types of storage facilities:

1) Baseload facility: The liquefaction facility and the storage tanks are located close to the source of the natural gas and are characterized by:

   - short residence time of liquid in the tank,
   - relatively constant LNG composition,
   - high fill rate from liquefaction plant, and
   - low nitrogen concentration in LNG.

2) Peak shaving facility: The liquefaction facility and the storage tanks are located in areas where there is a large seasonal variation in natural gas demand

and are characterized by:

- o long residence time of liquid in the tank,
- o relatively constant LNG composition,
- o low fill rate from liquefaction plant, and
- o high or low nitrogen concentration in LNG.

3) Receiving terminals: These tanks are usually located close to where LNG from tankers are normally unloaded. These tanks may also be used for peak shaving purposes or for accepting LNG in trailers. The characteristics of these tanks are:

- o short or long residence times
- o varying LNG composition
- o high or low fill rate from tankers or liquefaction plants or trailers
- o high or low nitrogen concentration in LNG

The computer model incorporates all the requirements listed above. As mentioned earlier, a reliable model of roll-over requires accurate calculation of physical properties. The ones that have to accurately be modeled over a wide range of composition are density, viscosity, heat capacity, thermal conductivity, saturation temperature, latent heat of vaporization, and the equilibrium constants of each component. Summarizing, the computer model has to accommodate:

- o LNG of at least three components and over a wide composition range
- o accurate physical properties over this range
- o accurate equilibrium properties over this range
- o short calculation time interval (0.01-0.05 hours)
- o extended simulation period (over 2000 hours)
- o variable or zero feed rate
- o variable feed type (tanker, liquefaction plant, trailers)
- o variable pressure, and
- o a simple, self-contained program that can be readily used.

Computer Model

The simulation model is similar to that described in other publications.[3] It begins with a tank of known dimensions, heat lead characteristics and containing liquid of known physical properties. The model considers either an initially stratified tank or a tank where stratification occurs as filling progresses. The program takes a certain interval of time (0.01-0.05 hours) and carries out a heat and component mass balance for each stratified layer. In every layer except the top layer the difference between heat input and heat output is the heat accumulated in that layer; on the top layer an iterative flash calculation is carried out since a part of the heat input is dissipated in the vaporization of liquid. This calculation procedure is continued until there is only one homogeneous layer in the tank.

ROLL-OVER: CALCULATION OF PHYSICAL PROPERTIES

For the two-component (methane-ethane) system, any physical property, F, of the system can be represented by

$$F = F_o + \Delta F$$

Where F_o is the value for pure methane and ΔF is the contribution of the second component. This ΔF is a function of the concentration of ethane X_1.

$$\Delta F = F_1 X_1$$

This is satisfactory for methane-ethane. However, for a three component mixture where one of the components is nitrogen, concentration X_2 the following equation may still apply:

$$F = F_o + F_1 X_1 + F_2 X_2$$

but F_1 and F_2 are not constants. Hence complexities has to be introduced in the roll-over model.

The initial easy solution was to use the physical properties calculated by the Air Products CAPP[7] correlations which are highly accurate but too complicated and large to be a part of a stand-alone, simple computer model for roll-over. Therefore another alternative was selected. It uses the same notation as before:

$$F = F_o + F_1 X_1 + F_2 X_2 + F_3 T$$

where T represents the temperature. For the correlations of density, heat capacity, thermal conductivity, and viscosity F_1, F_2 and F_3 are constants and independent of composition over a reasonably wide range of composition. These constants are obtained from the following correlations:

$$F_1 = (\frac{\partial F}{\partial X_1})_{X_2,T}$$

$$F_2 = (\frac{\partial F}{\partial X_2})_{X_1,T}$$

$$F_3 = (\frac{\partial F}{\partial X_3})_{X_1,X_2}$$

The correlations of the latent heat of vaporization, saturation temperature and equilibrium constants are more complicated and only the calculation of saturation temperature at a constant pressure is presented here for discussion.

Figure 5 represents qualitatively how the temperature varies with ethane and nitrogen respectively. The contribution of nitrogen is best represented by the term $C_1X_1 - C_2X_1{}^{C_4}$ where C_1 C_2 and C_4 are constants. Furthermore, while C_1 is independent of ethane, the second part, $C_2X_1{}^{C_4}$ is a function of the ethane concentration (Figure 6). Consequently, the final correlation takes the form:

$$T_{SAT} = T_0 + C_1X_1 + C_3X_3 + C_5X_1X_3 + C_2X_1{}^{C_4}$$

where T_0 is the saturation temperature of methane, C_1, C_2, C_3, C_4 and C_5 are constants, and X_1 and X_3 are nitrogen and ethane concentrations. The correlations for the equilibrium constants are more complicated but relatively easy for the computer to calculate. This capability of simple, yet accurate physical properties calculation allows the roll-over program to be an accurate, self contained package that can be run on small computers that do not have extensive library of physical properties of multicomponent mixtures.

ROLL-OVER: SIMULATION RESULTS

Table I compares physical properties of nitrogen-methane-ethane mixtures at 1 atm as calculated by Air Products' CAPP program and the correlations used in the roll-over program. Over an expected range of nitrogen and ethane concentrations the agreement is very good.

Figures 7, 8 and 9 each compare the results of the old roll-over program with the new one. Figures 7a, 8a, 9a represent results of the old model and 7b, 8b, 9b the new one. The three cases represent three known incidents of roll-over in LNG tanks containing no nitrogen. It is important to note that the modified program with improved physical properties correlations agree better with the observed results using two stratified layers. The old program used a three layer model; an intermediate layer forming during the initial filling stage had to be assumed to match the observed results. The three layer assumption has always been questioned since the increase in boil-off rates during mixing of the intermediate layer was never observed in the field.

Figure 10 represents the simulation results for a known case of roll-over where the LNG contained a substantial quantity of nitrogen. Figure 10a represents initial roll-over and 10b represents spontaneous roll-over. The agreements are excellent and whereas the old program required tedious updating of the physical properties from CAPP, the new program is self contained.

CONCLUSION

From the above results it can be seen that the new roll-over computer simulation program

- is a self contained package that can be readily used as an operational tool,
- has accurate physical properties correlation, and
- is better in simulating the actual roll-over incidents than the old program.

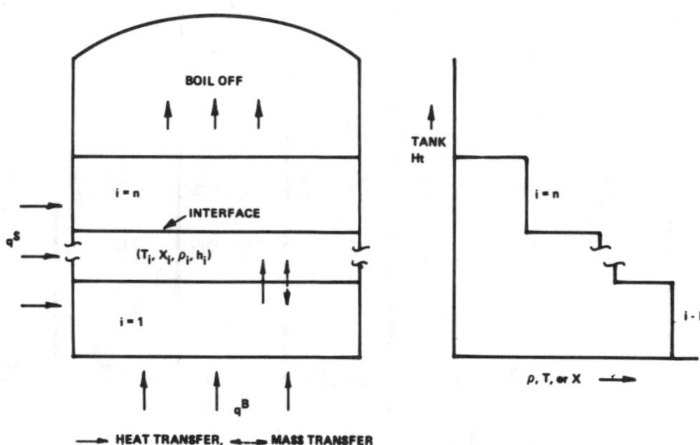

Figure 1. Schematic of a stratified storage tank.

LITERATURE CITED

1. Sarsten, J. A., "LNG Stratification and Roll-Over", (72-D-4), Presented at the A. G. A. Distribution Conference, Atlanta, Georgia, May 8-10, 1972.

2. Maher, J. B. and Van Gelder, L. R., "Understanding Roll-Over and Thermal Overfill in Flat Bottom LNG Tanks", (72-D-72), ibid.

3. Chatterjee N. and Geist, J. M., "Effect of Stratification on Boil-Off Rates in LNG Tanks", (72-D-37), ibid.

4. Private Communications, D. B. Crawford to J. M. Geist, April 19, 1972; Private Communications, J. M. Geist to D. B. Crawford, April 27, 1972.

5. Chatterjee, N., "Nitrogen Induced Stratification in LNG Storage Tanks", (76-T-69), Presented at the A. G. A. Transmission Conference, Las Vegas, Nevada, May 3-5, 1976.

6. Turner, S., International Journal Heat and Mass Transfer, 8, 759-767, 1965.

7. CAPP Program, Management Information Department, Air Products and Chemicals, Inc., Allentown, Pennsylvania.

TABLE I: COMPARISON OF PHYSICAL PROPERTIES CORRELATIONS

COMPOSITION (%)			DENSITY (LB/CU FT)		BUB. PT. (^oF)		HT. OF VAP. (BTU/LB)		EQUIL. CONSTANT NITROGEN		EQUIL. CONSTANT METHANE	
N_2	C_1	C_2	CAPP	MODEL	CAPP	MODEL	CAPP	MODEL	CAPP	MODEL	CAPP	MODEL
0.01	95	4.99	27.20	27.14	-257.7	-257.2	219.3	219.0	26.77	26.53	1.052	1.029
4.99	95	0.01	28.37	28.36	-279.8	-279.3	107.3	110.0	13.36	13.05	0.350	0.345
5	90	5	29.22	29.19	-280.4	-280.0	103.6	108.4	13.90	13.57	0.339	0.337
.01	90	9.9	27.98	27.98	-256.7	-256.2	218.9	218.2	29.05	28.50	1.111	1.085
9.9	90	.01	29.94	29.90	-291.3	-291.6	90.0	94.1	8.43	8.06	0.175	0.216
5	90	10	29.90	30.04	-280.3	-280.7	101.7	106.8	14.84	14.41	0.342	0.350
5	80	15	30.88	30.97	-282.4	-281.3	95.9	98.1	15.07	14.08	0.308	0.339
.01	80	19.99	29.44	29.56	-254.7	-254.1	218.2	216.6	34.81	32.64	1.249	1.231
.01	60	39.99	32.05	32.67	-249.7	-249.9	217.5	213.5	51.56	41.87	1.664	1.770
5	70	25	32.48	32.75	-285.2	-282.7	88.8	94.1	16.23	13.90	0.269	0.410

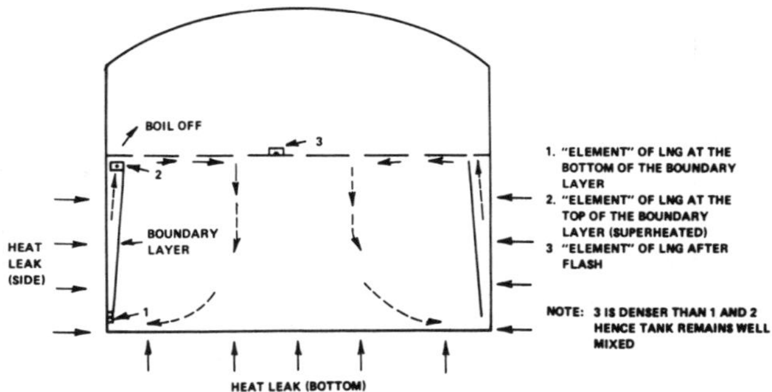

Figure 2. Circulation patterns in a storage tank containing LNG with no nitrogen.

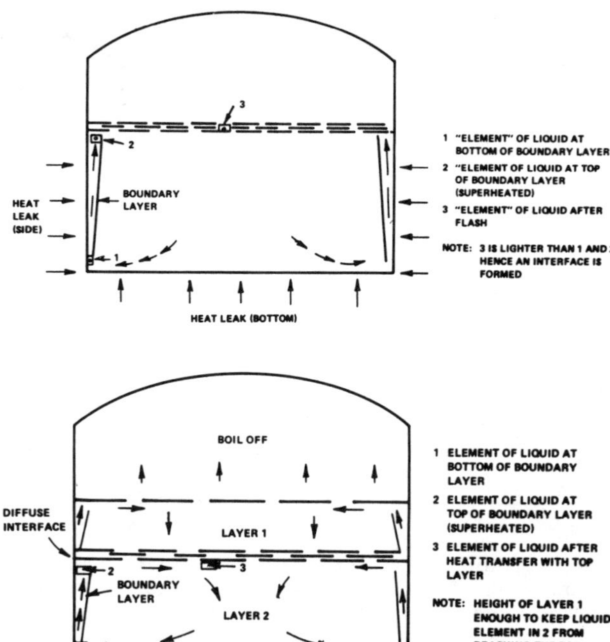

Figure 3. Mechanism of spontaneous stratification in a well mixed tank of LNG containing substantial concentrations of nitrogen.

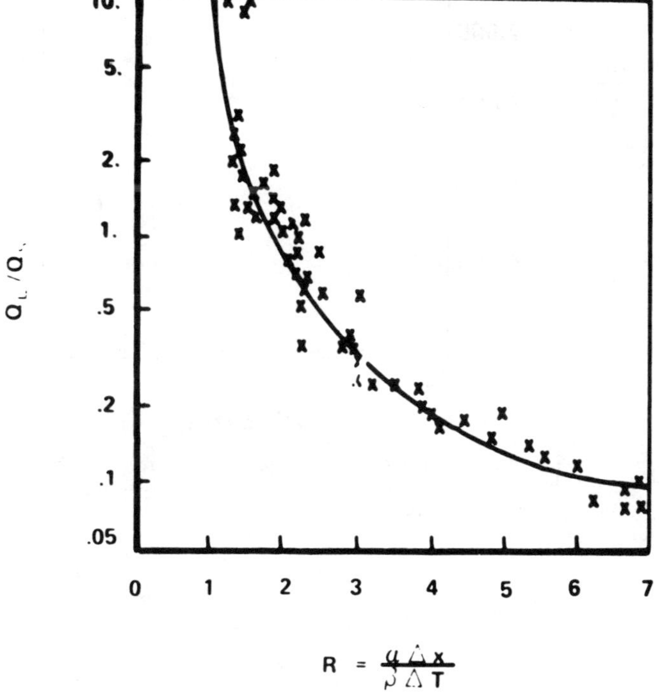

(a) THE RATIO OF HEAT FLUX ACROSS LIQUID INTERFACE TO THAT FROM A FLAT HORIZONTAL PLATE VS. R.

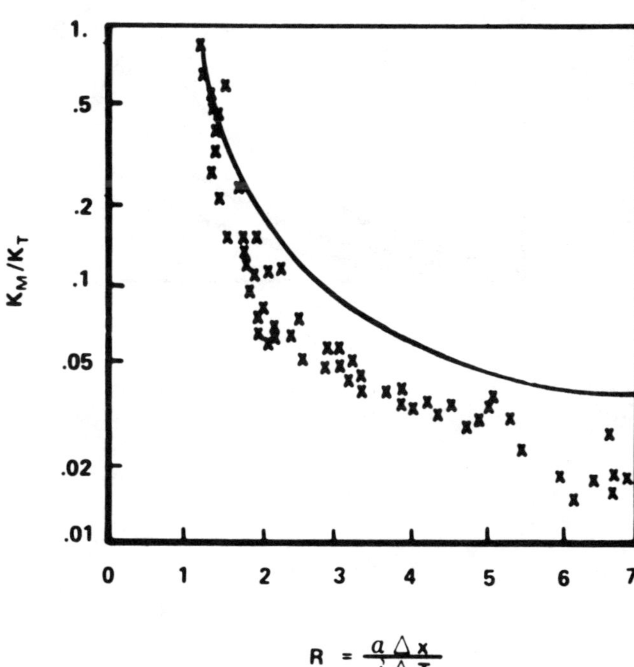

(b) THE RATIO OF MASS AND HEAT TRANSFER COEFFICIENTS ACROSS A LIQUID INTERFACE VS. R.

Figure 4. Heat and mass fluxes across a liquid-liquid interface vs. the stability parameter.

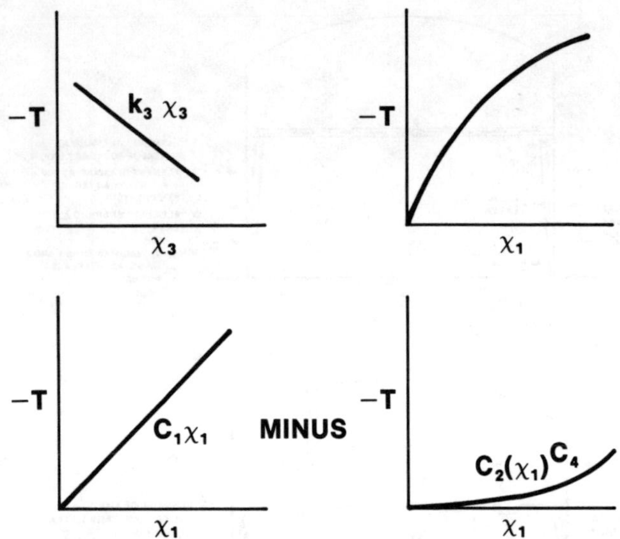

Figure 5. Contribution of ethane and nitrogen to LNG saturation temperature: (a) effect of ethane, (b) effect of nitrogen, (c) representation of (b) in a simple form.

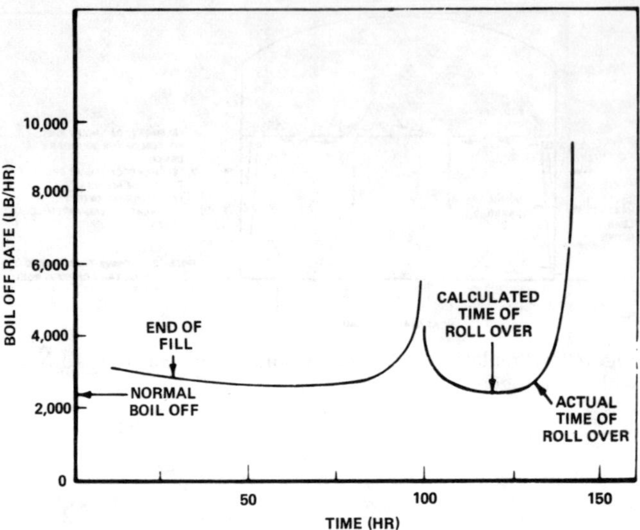

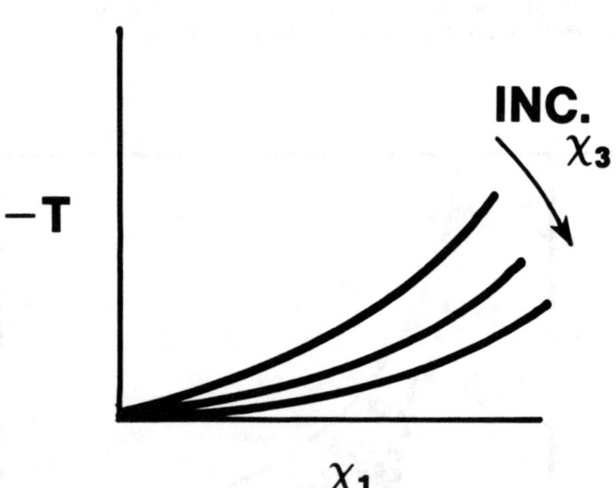

Figure 6. Final form of saturation temperature correlation: $T_{SAT} = T_o + C_1X_1 + C_3X_3 + C_5X_1X_3 + C_2X_1C_4$.

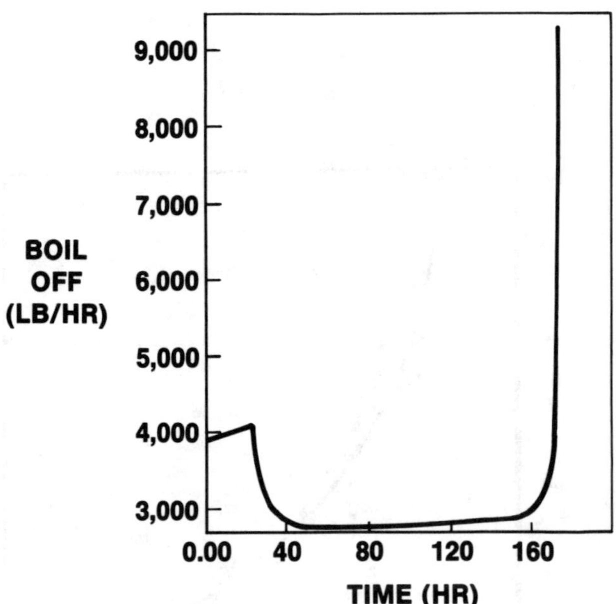

Figure 7. Comparison of old (a) and new (b) roll-over programs. Case I: high bottom fill rate, small quantity heavy liquid, large differences in temperature and density.

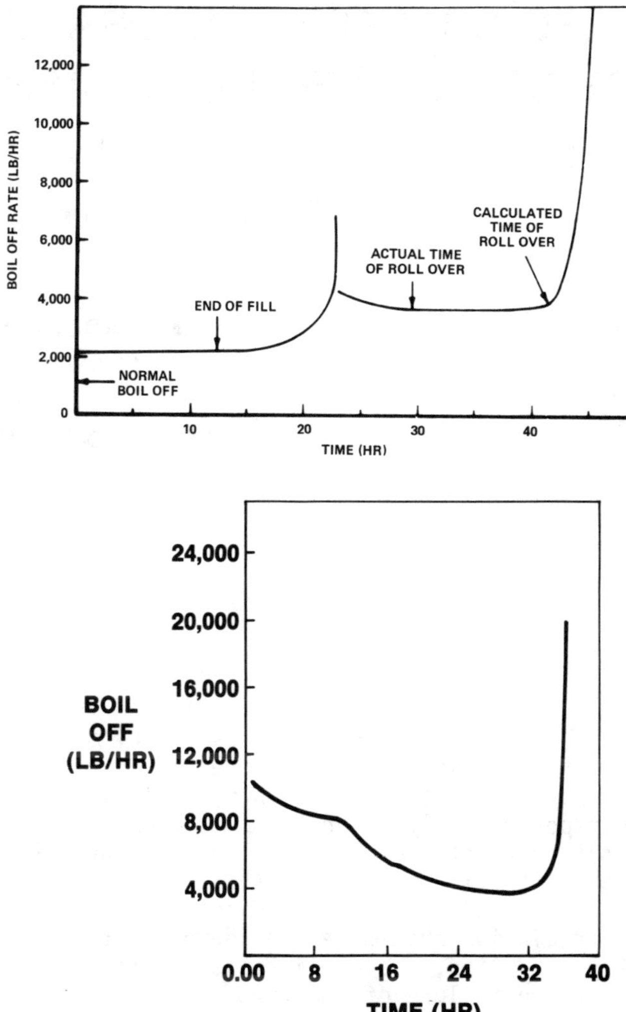

Figure 8. Comparison of old (a) and new (b) roll-over programs. Case II: high bottom fill rate, large quantity of heavy liquid, moderate temperature difference, small density difference.

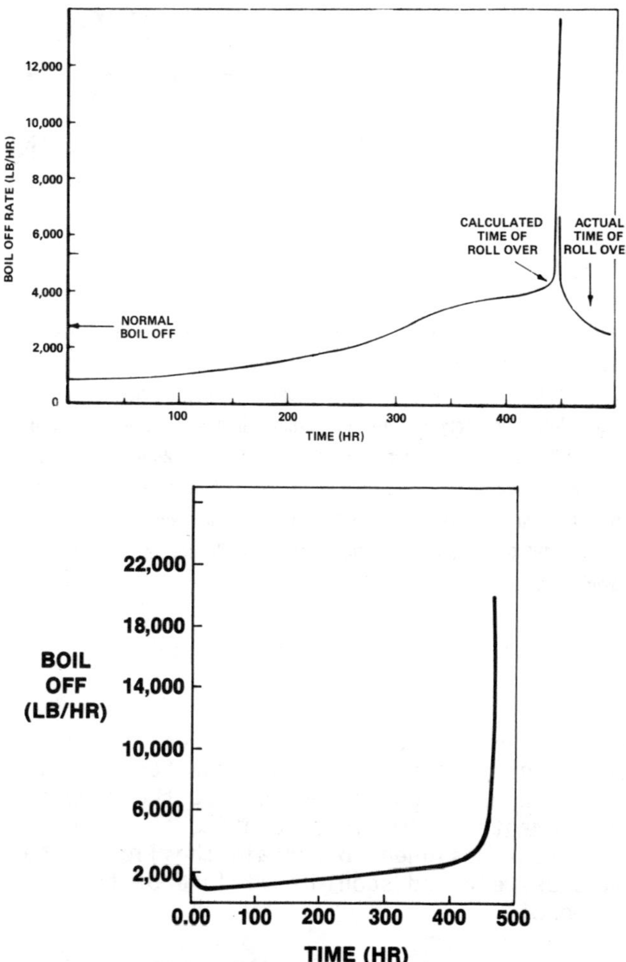

Figure 9. Comparison of old (a) and new (b) roll-over programs. Case III: low top fill rate, large quantity of bottom layer, small temperature difference, large density difference.

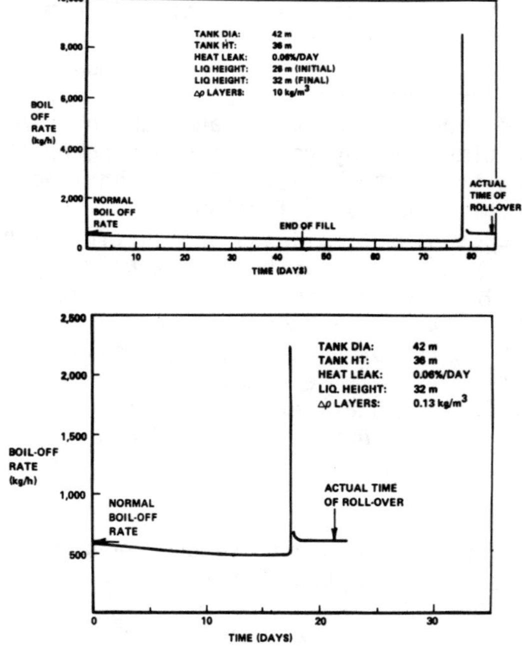

Figure 10. Results of roll-over using new program for LNG with nitrogen. (a) fill induced stratification and roll-over; (b) spontaneous stratification and roll-over.

DOUBLE-WALLED CRYOGENIC STORAGE TANKS—EFFECT OF PERLITE/FIBERGLASS INSULATION ON DYNAMIC LOADS IN CASE OF INNER TANK FAILURE

In double-walled LNG storage tanks the dynamic load on the wall of the outer tank in case of a sudden failure (zip-failure) of the inner tank can be significantly higher than the hydrostatic pressure. This was shown experimentally by Cuperus on a scale model. By using a simplified analytical model it is shown in this paper that insulation and gas containment in the annular space between the two tanks dampens this dynamic load. The dynamic to static load ratio in case of any empty insulation space can be reduced by a factor of two to three when the damping effects of the insulation is considered. The influence of insulation space dimensions, perlite stiffness, and the effect of fiberglass is shown in this paper.

A. S. ADORJAN
D. B. CRAWFORD
and
S. E. HANDMAN
M. W. Kellogg
Houston, Texas 77046

Concerns about catastrophic failure of the inner liquefied natural gas storage tank resulted in a new design concept called the double containment system, incorporating a concrete outer tank designed to contain the liquid that would be released should a failure of the inner tank occur.

Cuperus (1) discussed the technical details of various double containment storage systems, and he has proposed a reinforced concrete outer tank supported by an earthern berm. In another paper Cuperus (2) presented the results of scale model testing of sudden tank failure to derive dynamic loads on the outer tank. According to these results, substantiated by our own calculations, the peak of the dynamic load on the outer tank can be as much as six times the hydrostatic pressure on the tank at the bottom level. This high load and its asymmetrical distribution around the circumference of the wall was the justification for the need of an earthern berm to support the wall of the reinforced concrete outer tank. Cuperus, however, assumed and tested a model with an empty space between the shells of the two tanks. Thus, the damping effect of the insulation, such as perlite, and its load equalizing effect around the circumference of the wall was not accounted for.

A.S. Adorjan is now with Exxon Production Research Co., Houston, Texas

This study is considered as a very much simplified first step to explore the problem. More work needs to be done, both analytically and experimentally. For example, the effects of the gas contained between the two tank walls on the effective modulus of the insulation and the axial and/or tangential displacement of the insulation should be considered in the analysis. Also, the realistic simulation of the sudden failure of the inner tank is difficult to obtain in model tests because of scale effects. Also, recent and ongoing studies of the crack arrest properties of 9% nickel steel should be considered in evaluating the credibility of a catastrophic failure of the inner tank.

DISCUSSION

In order to make it easier to treat the problem by analysis, a simplified failure model is adopted. Figure 1 illustrates a typical LNG tank configuration. It is assumed that a crack propagates along the vertical length of the inner tank at a speed of about 0.4 times the velocity of sound in the tank material, which would take approximately 10 milliseconds based on the dimensions of Figure 1. Furthermore, it is assumed that the shell of the inner tank separates almost instantaneously from the flange of the bottom plate of the tank. While there is no experimental support for the latter assumption, it is a possibility which needs further verification. In any event, if the shell separates from the bottom plate, this would occur again very fast, similar to the

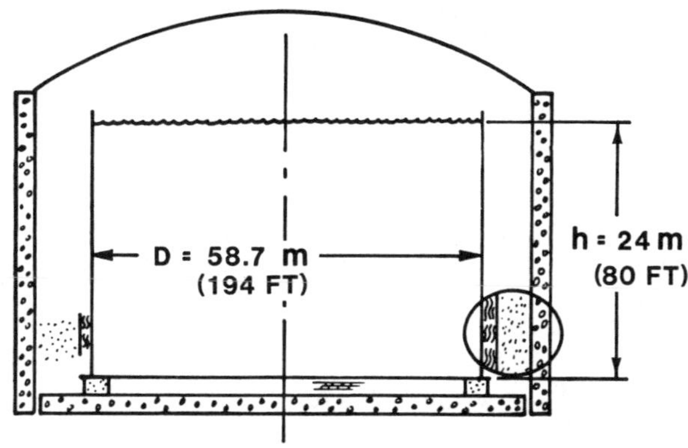

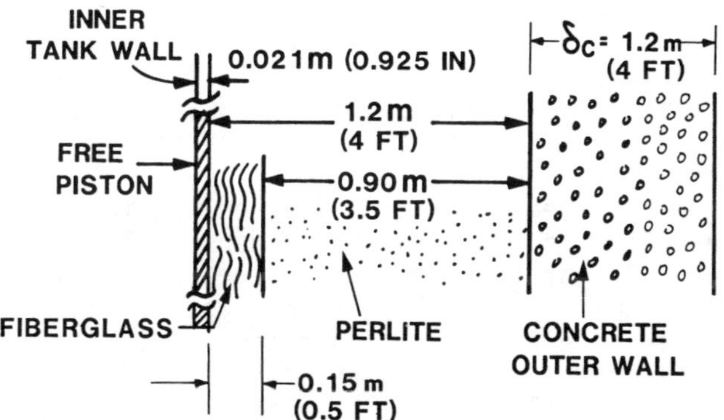

Figure 1. Tank dimensions (typical).

failure of the shell and at least an order of magnitude faster than the time required for the pressure to build up when perlite is present in the annular space between the inner and the outer tank as it is shown later. These assumptions reduce the problem to the "free piston" phenomenon. By selecting a surface element with unit surface area at the bottom of the wall of the inner tank, it can be visualized that after the failure of the inner tank the pressure of the fluid will accelerate the plate element. The accelerating force decreases as the velocity of the plate element approaches the theoretical velocity of the liquid, i.e. $v_o = \sqrt{2gh}$. In case the annular space contains an insulating material, the plate will initially accelerate, then it decelerates because of the resistance of the insulation, and it will come to rest when the available potential and kinetic energy of the plate is in equilibrium with the energy required to compress and/or crush the insulation. Figure 2 shows the geometry of the insulation space and plate movement. The space adjacent to the inner tank wall indicated by x_0 is filled with fiberglass.

In the interest of computational simplicity, the first part of this analysis assumes that the fiberglass is compressible with a negligible force, i.e. the modulus of elasticity is close to zero and the plate freely accelerates in this space without any resistance.

The force balance on a tank plate element with unit surface area and using the notations in Figure 2 can be written as:

$$m \frac{dv}{dt} = h\rho - \frac{\rho}{2g} v^2 - E \frac{x - x_o}{l - x_o} \qquad (1)$$

where the two terms on the right hand side represent the Bernoulli equation for the fluid behind the plate, and the third term is the resistance of the insulation, using the notations in Figure 2.

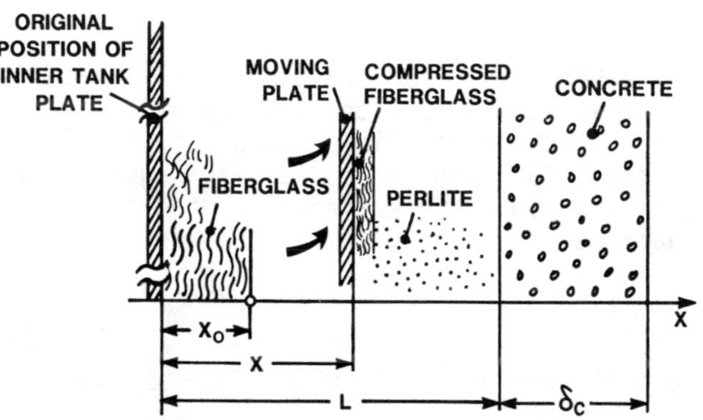

Figure 2. Notations for tank plate movement.

Equation (1) can be rearranged into a more suitable form for numerical solution:

$$\frac{d^2x}{dt^2} = \frac{h\rho}{m} - \frac{\rho}{2gm}\left(\frac{dx}{dt}\right)^2 - \frac{E}{m}\frac{(x - x_o)}{(l - x_o)} \qquad (2)$$

with the following boundary conditions:

$$t = 0 \qquad x = 0 \quad , \quad \frac{dx}{dt} = 0 \qquad (3)$$

It should be noted that for negative $x - x_0$ values, i.e. for the space filled with fiberglass, the third term of the equation representing the resistance (or restoring force) is equated to zero for the first part of the analysis.

The point of interest is where the velocity of the plate reaches zero, i.e. its energy is absorbed by the compression of the perlite. This also corresponds to the maximum excursion, x_{max} and restoring force. Then the factor of load amplification can be defined as the ratio of the restoring force and hydrostatic load, $h\rho$ at the bottom of the tank:

$$A = \frac{E}{h\rho}\left(\frac{x_{max} - x_o}{l - x_o}\right) \quad (4)$$

The modulus of elasticity of perlite insulation is taken from Novak (3) and Rakoczy (4). Novak performed the test to 7% strain and used an analytical procedure to compensate for friction on the wall of the test cylinder. A higher strain value is necessary for this study. Table 1 summarizes the results taken from both references.

Rakoczy (4) reported the stress-strain relationship for perlite up to about 57%. To account for the non linear behavior of this relationship which can be seen in Figure 3, the stress-strain curve was approximated by the equations

$$\sigma = 3.862\times10^{6}\,\epsilon \;,\; N/m^2 \qquad 0<\epsilon<0.35$$
$$\sigma = 1.425\times10^{6}\,\epsilon^{3.4128} + 95560 \qquad 0.35<\epsilon<0.70 \quad (5)$$

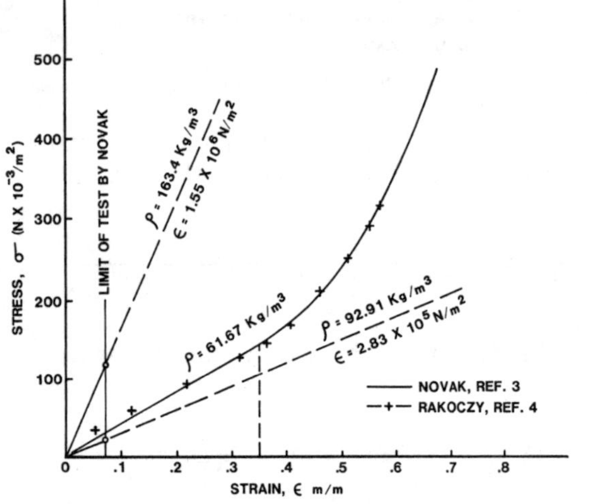

Figure 3. Stress-strain relationship for perlite.

The solution of Equation (2) is a harmonic motion in which the return leg is not valid because of the non-elastic behavior of the perlite.

Figure 4 gives the results of the solution of Equation (2).

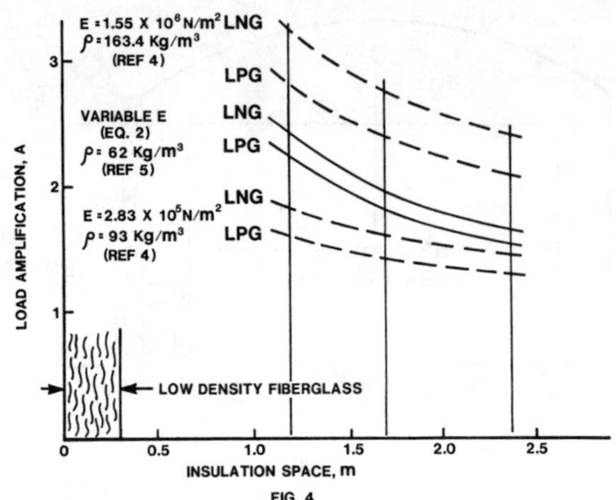

Figure 4. Dynamic load amplification.

Two sets of curves are given for each type of perlite insulation, one for LNG, another for LPG, where physical properties of liquid butane were used. In each case the "empty space" adjacent to the outside of the inner tank which is filled with low density fiberglass is kept to .3 m. The curves show that the dynamic load amplification factor decreases with increasing insulation space and increases with increasing modulus of elasticity of the perlite. Table 2 gives a typical example for the solution in tabulated form, corresponding to the E = 2.83×10^5 N/m^2 modulus of elasticity and an insulation space (distance between inner and outer tank) of 1.20 meters. Inner tank shell plate travel, velocity, acceleration, restoring force, dynamic load amplification and strain in the insulation is given in the table.

The curves for E = 0.283 and 1.55×10^6 N/m^2 modulus of elasticity are given for illustration to show the dependence of the amplification factor. Because the perlite compression tests were performed only to 7% strain while perlite strain values over 50% are needed for the analysis, the results are not accurate, and underpredict the amplification factor. The results are the most accurate for the 62 Kg/m^3 density.

Equation (2) was modified as:

$$\frac{d^2x}{dt^2} = \frac{h\rho}{m} - \frac{\rho}{2gm}\left(\frac{dx}{dt}\right)^2 - \frac{R(\epsilon)}{m} \quad (6)$$

where R(ϵ) is derived from Equation (5) with the consideration that in the fiberglass space the resistance is zero.

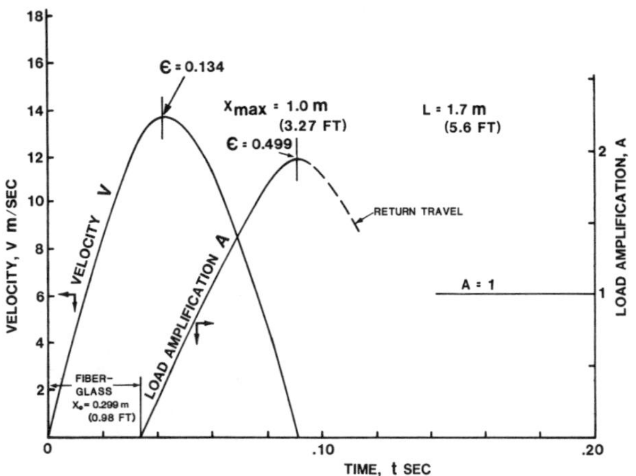

Figure 5. Dynamic load pulse.

Figure 5 shows a typical dynamic load pulse. The case corresponds to 62 Kg/m^3 perlite density, 1.70 m insulation space and LNG as fluid. The maximum amplification occurs at about 0.09 seconds, where the velocity of the plate element becomes zero. As it was formerly discussed, the return travel curve of the pulse is not valid because the analytical model does not account for hysteresis and crushing of the perlite; it will approach A = 1 (the static pressure condition) when the fluid comes to rest.

A maximum velocity of 17.07 m/sec is reached at about 0.042 seconds, significantly less than the maximum velocity obtainable if the plate accelerates without resistance, in the present case:

$$V_o = \sqrt{2gh} \qquad = 21.70 \text{ m/sec}$$

The limiting velocity in the insulation is the sonic velocity. For materials like perlite the sonic velocity can be defined as

$$c_o = (1 - \epsilon)\sqrt{\frac{g}{\rho_o}\frac{d\sigma}{d\epsilon}} \tag{7}$$

Using E = 2.83×10^3 N/m^2 and ρ = 93 Kg/m^3 the initial sonic velocity is 56.08 m/sec, significantly higher than the maximum velocity in Figure 5. This indicates that the insulation will homogeneously experience the effects of the inner wall movement. Evaluation of pressure gradients also indicates that the perlite can be expected to remain in position during the time frame of the zip failure.

Another factor influencing the amplification is the mass of the inner tank wall. Taking the E = 2.83×10^5 N/m^2 case at 1.20 meters insulation space, the amplification factor was determined for double and half wall thicknesses. For the double wall thickness A increased to 2.10 from 1.81 (16% increase), for half wall thickness A decreased to 1.52 (16% decrease).

Using the same conditions, E = 2.83×10^5 N/m^2 and 1.20 meters insulation space with 0.024 meters inner wall thickness, the fiberglass thickness (x_0) was varied. First it was decreased from .3 meters to half thickness, 0.15 meters. The amplification factor decreased from 1.81 to 1.68 (8% decrease). Decreasing the space to half thickness again, to 0.075 meters, A decreased to 1.50. For no fiberglass, A = 1.37.

These calculations show that the low modulus fiberglass is detrimental as far as the amplification factor is concerned. However, gas compression effect can substantially increase the effective modulus of the fiberglass. If the modulus of elasticity of the fiberglass is the same as the modulus of the perlite, the optimum condition can be obtained, that is, the lowest A as it was shown for the no fiberglass case.

CONCLUSIONS

By using a simplified analytical model, the load amplification factor in case of the sudden failure of the inner tank was determined. The load amplification factor was defined as the ratio of the dynamic load to the hydrostatic load on the outer tank wall at bottom level. The results indicate that the perlite in the annular space between the two tanks has a significant damping effect.

For a typical tank configuration with perlite insulation in the annular space the amplification factor is two to three times less than the factor of six obtained by Cuperus on scale model experiments with empty insulation space.

Table 1. Modulus of Elasticity of Perlite

Density Kg/m^3	Modulus of Elasticity (N/m^2)	Source	Limit of Test Strain, %
62.	3.86×10^5*	Rakoczy, Ref 4	55%
93.	2.83×10^5	Novak, Ref 3	7%
163.	1.55×10^6	Novak, Ref 3	7%

*approximately constant to about 30% strain

Table 2. Typical Computed Data

Input Data

w	=	182.4 Kg/m^2	Weight per area of plate
ρ	=	462.95 Kg/m^3	Fluid Density
E	=	$2.83 \times 10^5 N/m^2$	Perlite Modulus
h	=	24.00 m	Liquid Height
XO	=	0.3 m	Fiberglass Thickness
L	=	1.20 m	Insulation Space

Integration Step = .005 sec

TIME sec	DIST m	VEL m/sec	ACCEL m/sec^2	FORCE/AREA Kg/m^2	AMPLIF. A	STRAIN m/m
0.0	0.00	0.0	498.	0.0	0.0	0.0
0.005	0.01	2.97	587.	0.0	0.0	0.0
0.01	0.03	5.83	555.	0.0	0.0	0.0
0.02	0.11	10.88	448.	0.0	0.0	0.0
0.03	0.24	14.73	323.	0.0	0.0	0.0
0.04	0.40	17.25	185.	3344.2	0.3	0.11
0.045	0.49	17.53	28.	6141.3	0.55	0.21
0.05	0.58	17.08	-103.	8923.3	0.8	0.31
0.06	0.74	14.55	-294.	14040.5	1.26	0.49
0.07	0.86	10.55	-416.	18091.7	1.63	0.63
0.08	0.95	5.67	-497.	20709.5	1.86	0.72
0.09	0.98	0.26	-554.	21672.4	1.95	0.75
0.1	0.95	-5.48	-596.	20842.4	1.87	0.72

NOTATION

ϵ = Strain

ρ = density at zero strain

$R(\epsilon)$ = resistance of the insulation

E = modulus of elasticity of the perlite insulation

v = velocity of fluid

m = mass of inner tank wall plate per unit surface area

LITERATURE CITED

1. Cuperus, N.J., "Cryogenic Storage Facilities for LNG and NGL", 10th World Petroleum Congress, Paper No. 3, 1979.

2. Cuperus, N.J., "Developments in Cryogenic Storage Tanks", Sixth International Conference on Liquefied Natural Gas, Kyoto, Japan, 1980, Session II, Paper 13.

3. Novak, J.K., "Behavior of Perlite in Compression", Los Alamos Scientific Laboratory Report LA-DC-7158, 1965.

4. Rakoczy, E.J., "Load versus Deflection Test of Perlite", Pittsburgh-Des Moines Steel Company, Report 98391, Part VI, 1980.

THE CNG PROCESS: A NEW APPROACH TO PHYSICAL ABSORPTION ACID GAS REMOVAL

RALPH E. HISE
and
LESTER G. MASSEY

CNG Research Company
Cleveland, Ohio 44106

ROBERT J. ADLER
COLEMAN B. BROSILOW
and
NELSON C. GARDNER

Case Western Reserve University
Cleveland, Ohio 44106

WILLIAM R. BROWN
W. JEFFREY COOK
and
MICHAEL PETRIK

Helipump, Inc.
Brecksville, Ohio 44141

The CNG acid gas removal process embodies three novel features: 1) scrubbing with liquid carbon dioxide to remove all sulfurous molecules and other trace contaminants; 2) triple-point crystallization of carbon dioxide to concentrate sulfurous molecules and produce pure carbon dioxide; and 3) absorption of carbon dioxide with a slurry of solid carbon dioxide in organic carrier liquid. The CNG process is discussed and contrasted with existing acid gas removal technology as represented by the BENFIELD, RECTISOL and SELEXOL acid gas removal processes.

Acid gas removal from synthesis gas mixtures is a major step in many existing and proposed large-scale production processes such as ammonia, methanol, substitute natural gas, and synthetic liquid transportation fuel. We refer here to acid gas removal as the upgrading of crude gas streams by separation of carbon dioxide, hydrogen sulfide, and other substances usually present in small quantities such as COS, CS_2, HCN, and NH_3. Examples of large-scale production processes involving acid gas removal are the production of substitute natural gas and synthetic liquids from coal. The major steps are often 1) gasification, 2) shift conversion, 3) acid gas removal, and 4) catalytic reaction. Usually the acid gas removal step is the most costly of these four steps, both in capital cost and operating expense.

This paper describes a new acid gas removal philosophy and process of the physical type. Characteristics considered desirable by experts in gas purification are exhibited by the new process which is suited to pressurized gas streams (usually greater than 300 psia or 2,000 KPa) containing large amounts of carbon dioxide (usually exceeding 25%) plus some hydrogen sulfide, other sulfurous compounds, and trace impurities. Often these streams are rich in hydrogen and carbon monoxide.

In the following paragraphs, the characteristics desired in acid gas removal processes are reviewed. Then the new process is described, with particular emphasis on how the process philosophy achieves the desired characteristics. Several basic aspects are compared with Benfield, a hot carbonate chemical absorption process, and Selexol and Rectisol, two physical absorption acid gas removal processes. Finally, the present state of development of the process is outlined.

DESIRABLE PROCESS CHARACTERISTICS

Experts in the field agree that there are several characteristics desired in acid gas removal processes (1).

Ability to Remove Sulfurous Compounds and Trace Impurities

While H_2S is the main sulfur-containing species to be removed, there is also carbonyl sulfide (COS) and there are other sulfurous compounds present in trace quantities such as carbon disulfide, mercaptans, thioethers, and polysulfones. Sulfur-containing compounds pose a threat to most catalysts and are environmentally undesirable if rejected to the atmosphere with the carbon dioxide. Additional trace impurities include ammonia, hydrogen cyanide, and light hydrocarbons.

Hydrogen sulfide is removed to presently acceptable levels by all acid gas removal processes, but COS and other trace sulfurous compounds usually pose special problems. For example, COS often poisons chemical absorbents of the amine type by chemical reaction (2). Hot carbonate solutions absorb COS poorly because COS is not sufficiently acidic. Physical absorbents remove COS, but except for the new process of this paper, a greater absorbent flow rate is required for COS removal than for H_2S removal (3). In many designs, special hydrolysis reactors are added ahead of acid gas removal to convert the COS to H_2S, but complete conversion is problematical (1,2).

Ability to Produce Concentrated Hydrogen Sulfide

Hydrogen sulfide is normally converted to elemental sulfur in a Claus plant backed up by an off-gas treatment plant. The operability and cost of these plants depends on the concentration of the hydrogen sulfide feed. The minimum acceptable hydrogen sulfide feed concentration for Claus plant operability is about 25%; Claus and off-gas treatment costs drop sharply with increased concentration, and environmental pollution is also reduced. Claus feed concentrations of 40% hydrogen sulfide or greater are most desirable (4).

In many instances the acid gas ratio hydrogen sulfide:carbon dioxide is far below 1:3 in the crude gas stream to be purified by acid gas removal, so the separated acid gases must be concentrated with respect to hydrogen sulfide. In hot carbonate processes, the rate of hydrogen sulfide absorption is more rapid than the rate of carbon dioxide absorption, and this kinetic difference has been suggested as a basis for concentrating the hydrogen sulfide (7). The operability and flexibility of this operation may be limited. Physical absorption processes use a combination of flashing and reabsorption with lean solvent to concentrate the hydrogen sulfide. Because this means of concentrating hydrogen sulfide is costly, the separated acid gas stream from conventional physical absorption processes is typically 25% (or less) hydrogen sulfide.

Ability to Produce Pure Carbon Dioxide

Separated carbon dioxide is ultimately rejected to the atmosphere, so its purity, especially with respect to sulfurous species, is of environmental concern. Most acid gas removal processes have difficulty producing ultra-pure carbon dioxide, and there is generally extra cost associated with achieving high purity carbon dioxide.

Low Absorbent Flow Rates

The cost of acid gas removal is a strong function of absorbent liquid flow rate. High flow rates require large absorbers, heat exchangers, and stripping columns, large amounts of thermal energy for absorbent regeneration, and considerable mechanical energy for pumping regenerated absorbent. Low absorbent flow rates are desirable.

Low Energy Consumption

Thermal and pressure energy requirements for regeneration of acid gas absorbents can be especially large in chemical and conventional physical absorption acid gas removal processes. A desirable process would more closely approach the thermodynamically ideal reversible acid gas separation process to minimize energy costs.

CNG PROCESS DESCRIPTION

Consider a crude gas stream typical of the BCR Bi-Gas Process proposed over a decade ago for producing substitute natural gas from coal. The BCR Bi-Gas process was studied in detail by Air Products and Chemicals (5), and the crude gas composition, pressure, and flow rates of that study are used throughout this paper for illustrative purposes. The flow rate and composition of the crude gas stream (dry basis) are given in Table 1 (stream 305 (5)) for a 250MM SCFD substitute natural gas plant. Trace impurities including COS, HCN, NH_3, CS_2, CH_3SH, and light hydrocarbons are present, but are not included in Table 1. The pressure of the crude gas is 1,000 psia (68 atm or $6.9x10^3$ KPa).

Component	lb mole/hr	mole %
CO_2	32,406	32.45
CO	13,074	13.09
CH_4	12,192	12.21
H_2	40,536	40.60
N_2	516	0.52
H_2S	1,125	1.13
Totals	99,849	100.00

Table 1
Crude Gas Feed to Acid Gas Removal

A simplified CNG Acid Gas Removal Process flowsheet is shown in Figure 1. The crude gas stream defined in Table 1 enters at the lower left and passes through two heat exchangers which cool the gas to -56^oC, near the carbon

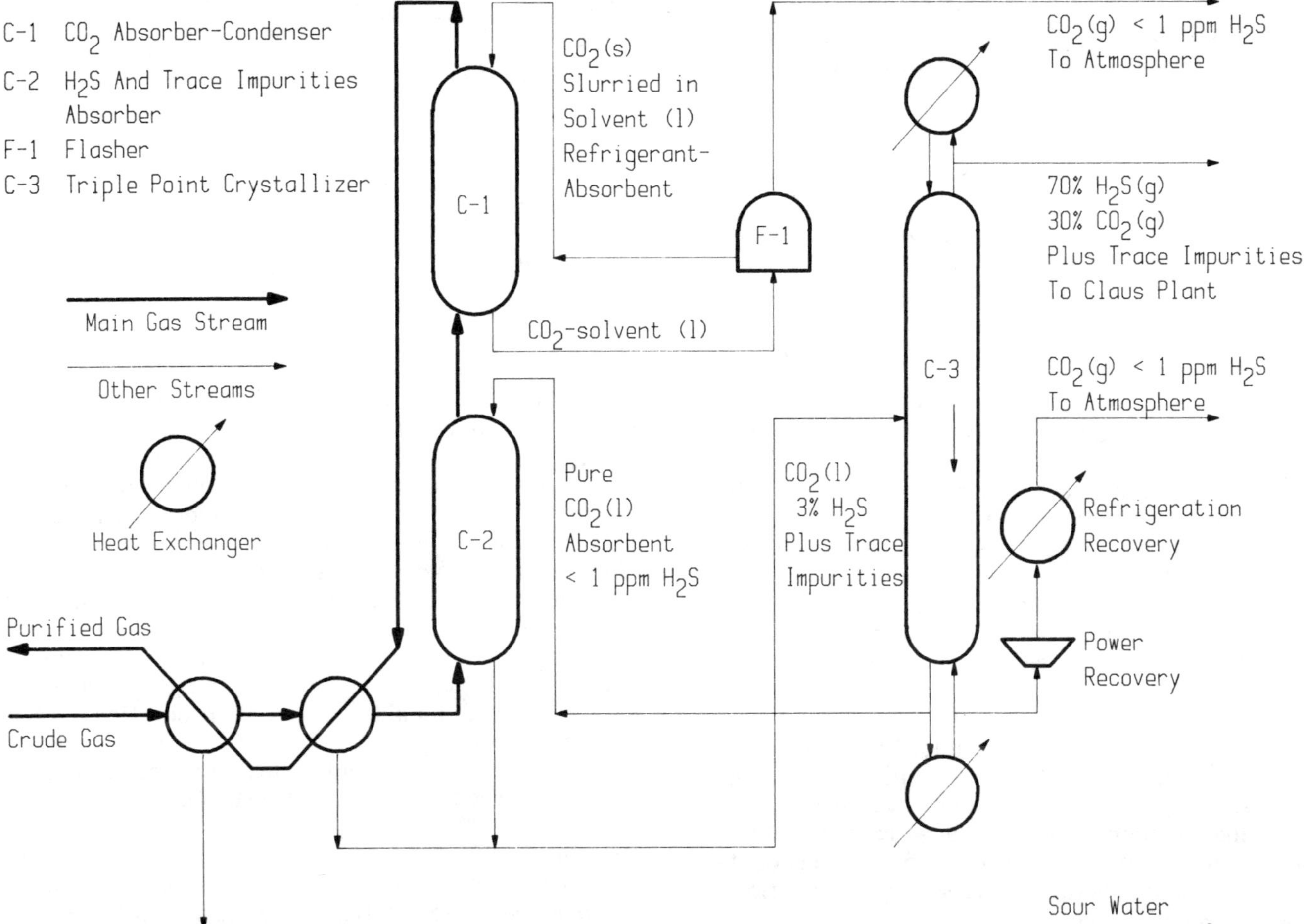

Figure 1. CNG acid gas removal process simplified flow diagram.

dioxide triple-point. Water is removed in the first heat exchanger and in a subsequent dehydration system (not shown). About 67% of the CO_2 in the crude gas stream is condensed in the second heat exchanger. These parts of the process are conventional.

The next process step is complete isothermal absorption of the H_2S, COS, and all other trace impurities from the main gas stream in absorber C-2 by a stream of pure liquid CO_2 absorbent. Pure liquid CO_2 has attractive properties as an absorbent: low molecular weight (44), low viscosity (about 0.35 cp), low surface tension (~10 dynes/cm), relatively high density (sp gr 1.18), and excellent absorption characteristics for COS and all trace impurities. To our knowledge, liquid carbon dioxide is the only physical absorbent which absorbs COS better than H_2S, i.e., $K_{COS} < K_{H_2S}$, where K_i is the equilibrium ratio of component i between phases ($K_i = y_i/x_i$, y_i = vapor mole fraction of component i, x_i = liquid mole fraction of component i).

The main gas stream, at -56°C, is now free of all sulfurous compounds and other trace impurities, but still contains about 14% CO_2. This CO_2 is removed by counter-current contact with a refrigerant-absorbent consisting of crystals of CO_2 slurried in a solution of CO_2 in an organic liquid. As CO_2 vapor is absorbed (condensed), the latent heat released melts solid CO_2. Thus, solid CO_2 circulated with the liquid provides direct refrigeration for the condensation and absorption of the CO_2 vapor, enabling a small absorbent flow to accommodate the considerable heat of condensation. The cold, fully purified main gas stream has its refrigeration potential recovered in two heat exchangers which cool the entering crude gas stream, and passes out of the acid gas removal process with less than 1 ppm of sulfurous compounds and other trace impurities, and 0.1 to 1.5% CO_2, depending on the application.

We now discuss the contaminated liquid CO_2 streams from the second heat exchanger and the H_2S absorber. These streams are combined and treated in a novel triple-point crystallizer which produces extremely pure CO_2 (<1 ppm impurities) and concentrated H_2S. Withdrawal of vapor (adiabatic flashing) from the top of the crystallizer C-3 cools the fluid and causes CO_2 crystals to form. The CO_2 crystals fall (sp gr 1.51) to the bottom of the crystallizer where they are melted by vaporized liquid withdrawn from the bottom of the crystallizer. Both liquid and vapor leave at the bottom of the triple-point crystallizer; pure CO_2 vapor is rejected to the atmosphere after recovery of power and refrigeration potential, and the pure CO_2 liquid is recycled as absorbent to the H_2S absorber. At the top of the crystallizer, a 70% H_2S acid gas stream is produced for feed to the Claus plant.

The organic liquid leaving the bottom of the CO_2 absorber-condenser C-1, now very rich with absorbed CO_2, is regenerated into slurry absorbent in F-1 by adiabatically flashing CO_2 gas to the atmosphere. As the CO_2 liquid vaporizes, it cools the remaining liquid and freezes some of the CO_2, thus regenerating the solid CO_2 slurry absorbent.

Figure 1 is a much simplified flowsheet. Multiple flashers, multiple crystallization stages, pumps, compressors, minor heat exchangers, and stripping columns for methane recovery have been omitted to show the basic process concepts clearly. More detailed descriptions and flowsheets are available (3,6).

CNG PROCESS CHARACTERISTICS

Several attractive characteristics are inherent in the CNG acid gas removal process.

Removal of All Sulfurous Compounds and Other Trace Impurities

Pure liquid CO_2 absorbs all sulfurous compounds and all other trace impurities at reasonable absorbent flow rates. Liquid CO_2 has the unique property, among all of the physical and chemical absorbents we have studied, that it absorbs COS better than it absorbs H_2S, as discussed previously in the section CNG PROCESS DESCRIPTION. Thus, an absorber designed to remove H_2S will automatically remove all COS and other trace impurities.

Production of Concentrated Hydrogen Sulfide

Triple-point crystallization can produce up to 75% H_2S concentrations without difficulty and with minimal energy consumption.

Production of Pure Carbon Dioxide

The triple-point crystallizer has exceptional ability to purify CO_2 since each stage of crystallization reduces impurities by factors ranging from 100 to 1,000. Present designs call for 1 ppm impurities, but 1 ppb or even less can be achieved with little additional capital or operating cost.

Low Flow Rates of Absorption Fluids

The flow rate of liquid carbon dioxide used to absorb sulfurous compounds and other trace impurities is less than the absorbent flow rate in any other physical absorption acid gas removal process we have studied. The low flow rate is partly due to low molecular weight and to the unique ability of liquid carbon dioxide to absorb carbonyl sulfide (and other trace impurities) better than it absorbs hydrogen sulfide. Low absorption temperature also contributes to the low flow rate; hydrogen sulfide is absorbed at -56°C,just slightly above the lowest temperature carbon dioxide can exist as liquid (triple-point temperature, -56.6^{o}C). Even at the triple-point, liquid carbon dioxide physical properties are attractive for an absorbent fluid -- low viscosity, low surface tension, and relatively high density (see CNG PROCESS DESCRIPTION).

Similarly, because of its enhanced heat capacity, the slurry used for carbon dioxide removal has a lower flow rate than conventional physical absorbents. Condensation of carbon dioxide vapor to liquid releases considerable latent heat. Dissipating this latent heat without excessive absorbent temperature rise in most absorbers requires interstage cooling with heat exchangers or very large liquid absorbent flow rates to provide sufficient heat capacity. The slurry absorbent provides the necessary heat capacity in situ and in a concentrated form through the nearly isothermal phase change solid carbon dioxide to liquid carbon dioxide.

Low Energy Consumption

The CNG acid gas removal process utilizes relatively small temperature and pressure driving forces to effect the desired separations: 1) bulk carbon dioxide from crude gas by condensation; 2) minor acid gases from crude gas by physical absorption; 3) carbon dioxide from hydrogen sulfide by triple-point crystallization; 4) carbon dioxide from crude gas by absorption with slurry absorbent; and 5) carbon dioxide from solvent by adiabatic flashing. The CNG process uses no absorption agent foreign to the crude gas to remove sulfurous compounds and trace contaminants, thus easing the subsequent recovery of absorption agent (carbon dioxide) from acid gases (carbon dioxide, hydrogen sulfide, trace contaminants). Hence the CNG process is more nearly reversible in the thermodynamic sense than are other acid gas removal processes, and the CNG process energy requirements are only a fraction of those normally experienced in acid gas removal.

PROCESS COMPARISONS

Table 2 compares some of the major characteristics of the CNG process with commercially available acid gas removal processes: Benfield (hot carbonate), Selexol (dimethyl ether of polyethylene glycol), and Rectisol (refrigerated methanol). Benfield is a chemical absorption process; Selexol and Rectisol are physical absorption processes. The comparisons are for the crude gas stream defined in Table 1.

The volumetric absorbent flow rate in Selexol and Rectisol for hydrogen sulfide removal is about twice that in the CNG process. If Selexol is required to remove all COS in addition to hydrogen sulfide, the flow rate of Selexol solvent is about eight times the flow rate in the CNG process. The lower CNG absorbent flow rate is a result of: 1) unique COS absorption capability of liquid carbon dioxide; 2) lower molecular weight of carbon dioxide (44) as compared to Selexol solvent (about 200); 3) lower absorption temperature possible in the CNG process. When compared with Rectisol, the lower temperature of absorption in the CNG process is beneficial.

The volumetric absorbent flow rate for carbon dioxide removal in the CNG process is 1/6 to 1/30 of that for Selexol and Rectisol. This lower flow rate is a result of: 1) lower temperature of absorption, and 2) prior bulk condensation of carbon dioxide. The high effective heat capacity of the slurry containing solid carbon dioxide (an in situ phase change refrigerant) is highly beneficial.

Thermal energy needed to regenerate the Benfield hot carbonate solution is high because it is a chemical absorbent; that needed by Selexol and Rectisol is large because the absorbent flow rates are large. In the CNG process, liquid carbon dioxide absorbent used for hydrogen sulfide and other trace impurity removal is regenerated by energy-efficient triple-point crystallization. The slurry absorbent is regenerated by simple adiabatic flashing.

High hydrogen sulfide concentration (up to 70% H_2S), achieved by triple-point crystallization in the CNG process, has most desirable economic

Table 2
CHARACTERISTICS OF ACID GAS REMOVAL PROCESSES (1)

Process	Type	~T°C	Absorbent, MW	Gal/Hr to Remove H_2S	Gal/Hr to Remove CO_2	Regen Steam, lb/hr	Claus Feed % H_2S	CO_2 to ATM., PPM S
CNG	PHYS	-56	CO_2L,S 44 (4)	60,000 (8)	50,000 (10)	0	70	< 1
BENFIELD (2)	CHEM	100 +	K_2CO_3 ~(5)	-	-	1,300,000	15-20	3,400
SELEXOL (3)	PHYS	25	DEPG 200 (6)	120,000 480,000 (9)	1,200,000	140,000	25	100
RECTISOL (3)	PHYS	-40	CH_3OH 32 (7)	100,000 150,000 (9)	1,500,000	100,000	25	100

(1) Based on BIGAS process, 250 MMSCFD of SNG, ref. [5].
(2) See reference [5].
(3) Estimated from reference [7].
(4) CO_2 liquid; CO_2 solid in liquid.
(5) Hot solution in water.
(6) Dimethyl ether of polyethylene glycol.
(7) Refrigerated methanol.
(8) Liquid CO_2 absorbent.
(9) Upper figure to remove COS.
(10) CO_2 solid in liquid absorbent.

consequences in sulfur recovery via the Claus process. Other available acid gas removal processes have difficulty producing sulfur recovery feed streams with more than 25% H_2S, thus incurring added costs for handling the diluent gases.

The carbon dioxide discharged to the atmosphere is extremely pure in the CNG process, again a consequence primarily of triple-point crystallization. If desired, food-grade carbon dioxide can be produced with virtually no change in process economics.

PRESENT STATUS OF CNG PROCESS

Research on the new acid gas removal process has been ongoing for the past seven years by the CNG Research Company (a wholly owned subsidiary of the Consolidated Natural Gas Company), and for the past two years, in cooperation with the United States Department of Energy (Morgantown Energy Technology Center, Contract No. DE-AC21-80MC14399). Most of the research has been done by Helipump Corporation at Case Institute of Technology, Case Western Reserve University, Cleveland, Ohio.

To date the work has encompassed process conception, flowsheet preparation, economic analysis, experiments to develop an extensive data base of equilibrium phenomena and critical rates. A small process development unit has been constructed and operated to obtain triple- point crystallization data. A second process development unit has been constructed and operated to observe slurry formation and handling, slurry pumping, and qualitative slurry absorption characteristics. Current efforts are directed to the construction of a larger, well-instrumented, carbon dioxide triple-point crystallizer.

The CNG process is now in the latter stages of development; commercial applications are anticipated toward the end of 1985.

LITERATURE CITED

1. Discussions and Proceedings of the "Conference on Gas Cleaning in the Production of Synfuels," sponsored by the Chemical Engineering Department, North Carolina State University, Raleigh, N.C., April 29-30, 1982.

2. Eickmeyer, A.G., and H.A. Gangriwala, "The Role of Acid Gas Removal in Synfuels Production," presented at the Symposium on Gas Purification at the AIChE National Meeting, Houston, Texas, April 5-9, 1981.

3. Adler, R.J., et al., "The CNG Process, A New, Low Temperature, Energy Efficient Acid Gas Removal Process," presented at the Symposium on Industrial Gas Separations, American Chemical Society, Washington, D.C., June 14-15, 1982.

4. Beavon, D.K., Kouzel, B., and J.W. Ward, "Claus Processing for Novel Acid Gas Streams," Symposium on Sulfur Recovery and Utilization, presented before the Division of Petroleum Chemistry, Inc., American Chemical Society, Atlanta Meeting, March 29-April 3, 1981.

5. "Engineering Study and Technical Evaluation of the Bituminous Coal Research, Inc. Two-Stage Super Pressure Gasification Process," Air Products and Chemicals, Inc., Contract No. 14-32-0001-1204, Research and Development Report No. 60, prepared for the OFFICE OF COAL RESEARCH.

6. Adler, R.J., Brosilow, C.B., Brown, W.R., and Gardner, N.C., "Gas Separation Process," U.S. Patent 4,270,937, June 2, 1981.

7. Kohl and Riesenfeld, Gas Purification, 2nd Ed., Gulf Publishing, 1974.

8. Christensen, K.G., and W.J. Stupin, "Comparison of Acid Gas Removal Processes," U.S. Department of Energy Report Fe-2240-49, prepared by C.F. Braun & Co., under Contract No. EX-76-C-01-2240, April, 1978.

9. Edwards, M.C., "H_2S-Removal Processes for Low-BTU Coal Gas," U.S. Department of Energy Report ORNL/TM-6077, Distribution Category UC-90c, January, 1979.

ABSORPTION REFRIGERATION IN CRYOGENIC SERVICES

Absorption refrigeration is a means for energy conservation via recovery of low-level heat energy. It can yield energy cost reductions in industrial operations when low-value by-product heat is available. While an absorption refrigeration cycle may be operated at cryogenic temperatures, large-scale commercial experience is limited to −40° and above. This chemical cycle is investigated here as a viable alternate to the high-end stage of conventional cryogenic cascades. Three examples compare aqua-ammonia absorption refrigeration (AAR) with mechanical (vapor compression) refrigeration in important commercial cryogenic operations:

- **Liquefaction of natural gas for storage and transport.**
- **Production of ethylene by the pyrolysis of hydrocarbons.**
- **Production of oxygen in air-separation plants.**

The savings gained by the proposed substitution of absorption refrigeration accrue from trading low-value heat energy for high-cost work energy. They depend on the availability, nature and optimum utilization of "waste" or by-product heat and thus vary widely from case to case.

MARCEL J. P. BOGART

Fluor Engineers, Inc.
Irvine, California

INTRODUCTION

Absorption refrigeration operates mainly on a heat energy input. Mechanical refrigeration requires work energy inputs, which almost invariably involves the consumption of fossil fuels. The debilitating, ever-escalating cost of these fuels creates opportunities for economic enhancement by reverting to absorption refrigeration.

Fig. 1 is a simplified flow diagram of an AAR unit. Here the refrigerant ammonia vaporized in the evaporator is not compressed, but is absorbed in water or in "weak-aqua" - a lean aqueous ammonia solution. The cycle is completed by separating the resulting absorbate ("strong-aqua") by fractional distillation. The flow diagram may become considerably more complicated when accommodating such variants as multiple refrigeration loads and levels, multiple heat input sources, and opportunities for internal heat recuperation.

AAR operations are reported as feasible down to -60°C (-76°F). However, the decrease in efficiency with temperature and the penalties of vacuum operations suggest limiting this large-scale AAR application study to its replacement for the high-end (-40° and above) of a mechanical cascade using multiple or mixed refrigerants.

The author is presently a consultant at
10602 Cordoba Court
Whittier, CA 90601.

First-round process designs for three cases (defined below) were made using published procedures (1) to compare their relative energy expenditures. Optimization and payout estimates are not included as they depend largely on the characteristics of the heat source(s) available, which vary widely from one application to another. Given the specific project parameters, the value of the energy saved may be calculated from the data presented in this report.

CASE A - LNG Plants

Liquefied natural gas plants permit the transportation of this premium fuel from remote regions (the Arabian peninsula, Malaysia, Alaska's North Slope, etc.) to distant major markets. The liquefaction operation is illustrative of a single large refrigeration load (at about -162°C/-260°F) with little or no on-site recovery of the cold produced.

Such plants also separate and recover components other than methane from the natural or associated gas feedstock. The extent of these facilities vary greatly, given the wide spectrum of feedstock composition. The refrigeration system and requirements are thus complex. For the present exploratory purpose they are simplified to the single-load duty of condensing prechilled pure methane gas at -162°C (-260°F). Setting the module size at liquefying 100 million SCFD (110,600 Nm^3/h) gives a refrigeration duty of 40.56 GJ/h (38,440,000 Btu/h).

As further simplification, we will not consider the mixed-refrigerant options, but will base our comparison on the conventional compression cascade, using pure methane, ethylene, and propylene as refrigerants. Fig. 2 gives the work energy consumed by such a cascade as a function of the base load temperature. In this case, AAR will substitute for the propylene section of the cascade and will save 18,430 kW (24,710 HP). The AAR duty will then be a single-load at -40°; cascade calculations indicate that it must condense 389,000 kg/h of superheated ethylene vapor from 1884 kPa/-10°C to liquid at 1849 kPa/ -31°C. Neglecting losses, the corresponding duty is 145.8 GJ/h (138,200,000 Btu/h).

CASE B - Ethylene Plants

Ethylene and propylene are produced by the pyrolylysis of hydrocarbons (ethane through heavy oil). The cracked gas contains hydrogen and methane, which are removed by condensation and distillation at temperatures ranging to below -100°C (-148°F). Here refrigeration is also used at diverse levels in the condensation and distillation steps which recover and purify the light-olefin products. The plant refrigeration system is complicated by the number and variety of the loads and by opportunities for cold recovery. This example is in strong contrast to the single-load LNG plant concept.

Our purpose can again be served by simplifying this system to the cooling down of the pyrolysis gas, using the data of Picciotti (7) for the production of 600,000 metric tonnes of ethylene per year, spanning the range of 15° to -122°C (59° to -188°F) in a multiple-refrigerant cascade. Again, we will substitute AAR for Picciotti's high-stage (propylene) cycle, which cools the gas from 15°C (59°F) to -35°C (-31°F). The system efficiency can be maximized by subdividing the load into separate temperature levels. The refrigerant temperature and duties of the three-level cascade used here are:

Temperature	Refrigeration Duties
1. -40°	13.5 GJ/h (12.8x10^6 Btu/h)
2. -25°C (-13°F)	21.7 " (20.6 " ")
3. 5°C (23°F)	22.6 " (21.4 " ")
Total	57.8 " 54.8 " "

The work energy reduction due to the high-end AAR substitution is 5925 kW (7945 HP) in this case.

CASE C - Oxygen Plants

In the typical oxygen plant, the air feed stream is chilled to fractionation temperatures by heat exchange with the oxygen and nitrogen product gases. The refrigeration requirement (for the irreversible work of separation and inevitable heat and approach losses) is supplied to an internal expansion cycle as additional energy input to the feed air compressor. In an application of interest today, the gaseous oxygen product may be compressed for delivery to a pressurized coal gasification unit. At first glance, such a plant does not provide an opportunity for economies by using the AAR technique.

However, the energy consumed in a gas compressor is proportional to the absolute temperature of the suction gas. Refrigerating this gas from near-ambient to, say, 5°C will reduce the work energy expended by about 10%. If low-cost waste heat is available, the AAR unit will save the fuel-cost equivalent of the work energy reduction at little more than the amortization on the added capital investment.

The basis for this case is the production of 2,000 metric tonnes per day of contained oxygen at 98% purity, compressed in three stages to a delivery pressure of 3,600 kPa (523 psia). The nitrogen is vented. The intake air is at 35°C (95°F) with a 23°C (73°F) wet-bulb temperature, and is compressed in two stages to 542 kPa (79 psia). As shown in Fig. 3, a spray cooler is used on the intake air to minimize horsepower loss from excessive suction pressure drop in the compressor. The AAR unit refrigerates the suction flow to each stage, with a minimum refrigerant temperature of 2°C (36°F) to avoid icing problems. The refrigeration scheme for the oxygen compressor is the same as that for the air compressor in Fig. 3, except that three stages of compression are used.

Gas compression horsepower calculations (excluding mechanical and drive-train losses) give:

Operation	Compressor kW Air	Oxygen
Conventional	25,080	10,240
Refrigerated	22,930	9,280
Savings	2,150	960
Total	3,110 (4170 HP)	

A three-level cascade providing the necessary refrigeration is subdivided thusly:

	Temperature	Refrigeration Duty				
1.	2°C(35.6°F)	14.07 GJ/h	(13.34x10^6	Btu/h)		
2.	9°C(48.2°F)	21.83	"	(20.69	"	")
3.	17°C(62.6°F)	7.66	"	(7.26	"	")
	Total	43.56	"	(41.29	"	")

AMMONIA ABSORPTION REFRIGERATION

The actual design of the AAR units for each of the above-defined cases are discussed below by sections:

EVAPORATION: The Case A refrigerant evaporation cascade, Fig. 4, is typical of all three cases. Advantage has been taken here of the very limited opportunity for efficiency enhancement by cascading. The refrigerant (100 ppm H_2O) is prechilled by cold recovery exchangers on the evaporator vapors and by a high-pressure flash (207 kPa/ 30 psia). The intermediate level evaporator partially desuperheats the compressed ethylene gas hot medium (ca. 7% of the total duty). Blowdown of the low-level evaporator liquid guards against water accumulation problems (2, 5). The blowdown rate is 0.41% of the refrigerant supplied; its composition is 1.22%w H_2O.

The Case B evaporator cascade provides, as noted, refrigeration services at all three levels. Its design differs mainly in the use of a separate, higher-purity refrigerant (10 ppm H_2O) in the lowest level. The Case B blowdown flows are:

Refrigerant ppm(wt) H_2O	Blowdown rate	Blowdown composition
10	0.10%	0.17% H_2O
500	0.85%	4.60 "

The Case C evaporator cascade is again different from that in Fig. 4 in having evaporators at all three levels. The blowdown rate in this case is 1.5% of the supply; it contains 1.63%w H_2O.

ABSORPTION: If proper care is taken to keep the AAR unit free of non-condensible gases, there is no need for the classical countercurrent absorber, whose function is to separate highly soluble gases from low-solubility ones. The task of an AAR "absorber" is simply to dissolve ammonia vapor in water. This involves mainly removing the very large heat of solution and is best done in a properly-designed heat exchanger (5). Feeding a well-mixed blend of evaporator vapor and liquid absorbent to such an absorber-condenser gives a two-phase inlet mixture whose temperature may be 10° to 20°C higher than that of either feed stream due to the heat of solution released. The operation is conducted to give complete condensation of the throughput mixture, with the liquid absorbate as the sole effluent.

Fig. 4 also shows the absorption cascade for Case A. As seen, the absorbent flows in series from one level to the next higher one, becoming enriched in this case from 16.1%w NH_3 in the weak-aqua to 26.2%w NH_3 in the strong-aqua feeding the fractionator. In this case 82% of the total vaporization in the three evaporator levels emanates from the lowest level. This allows the simplification of having only one absorber-condenser (E-6, Fig. 4) for removing the heat of absorption of ammonia in water. The small heat release from absorbing the lesser quantities of ammonia vapor flowing to the mid- and high-level absorbers gives a tolerable temperature rise to the absorbent stream without the further use of air-or water-cooled exchangers.

In the alternate (parallel) arrangement where weak-aqua absorbent is fed separately to each stage, absorber-condensers will be required at all levels. Each absorbate may be fed separately to its appropriate entry-point in the fractionator. It is doubtful that the small gain in fractionation efficiency will justify the added complications.

The Case B absorption cascade has a more even distribution of evaporator vapors from the three levels. Absorber-condensers are required at each level. Also, as will be described, additional liquid and vapor streams from the fractionation section are fed to this absorption cascade. This cascade is also supplied with a 16.1%w NH_3 weak-aqua; it sends a strong-aqua to the fractionator analyzing 36.2%w NH_3.

The Case C evaporation is also more evenly balanced. Its absorption cascade thus has absorber-condensers at each level. It boosts the 22.5%w NH_3 inlet weak-aqua to a 58.1% strong-aqua fractionator feed.

The temperature rise in the low-level mixer results in an absorber-condenser inlet temperature of 71°C (159°F). The condensing-cooling duty of this absorption step is therefore divided between two exchangers in series. The first one provides part of the fractionator feed preheat; condensation of its ammonia-water two-phase effluent is finished in a second, water-cooled exchanger. The heat consumption of this AAR unit is reduced by over 15% by this internal heat-reclamation scheme.

FRACTIONATION: In Fig. 1 the separation of ammonia from water appears to be simple enough for an undergraduate chemical engineering exercise. However, it is here that the external supply of heat energy is consumed. The mandatory maximization of its efficiency thus calls for skill and expertise in its design and in matching its demands to the available heat source(s) and coolant(s). The various energy-saving techniques employed in these designs are shown in Fig. 5.

The first of these is the condensation of the fractionator reflux in a partial condenser. The large departure from ideality of the highly polar ammonia-water system causes a large decrease in the requisite reflux ratio with the water content of the overhead product (3), even in the narrow range of nearly-anhydrous ammonia down to 99% NH_3. Here the composition ratio of water in the reflux drum liquid to that of its equilibrium vapor is in the range of 400 to 500. Obtaining the overhead product as a vapor from the reflux drum (sent to a final condenser) thus provides important energy savings at a slight increase in equipment cost.

The reflux drum liquid and vapor compositions and the reflux ratios for the three cases are:

Case	Liquid wt% NH_3	Vapor ppm H_2O	Reflux Ratio
A	96.70	100	0.1162
B	99.66	10	0.1680
C	91.04	300	0.1016

These vapor compositions assume that the mechanical design of the reflux drum limits its liquid entrainment rate to 0.1% of the effluent vapor.

In Cases A and B a further increase in distillation efficiency is obtained by doing part (71% and 31%, resp.) of the condensing duty in an 'intermediate condenser' located at a lower point in the column. Here the condensing temperature is high enough to reclaim the heat released as a part of the fractionator feed preheat. The relatively high ammonia content of the Case C feed indicates that the added complication may not be cost-effective in this instance.

The ubiquitous feed/bottoms exchanger is also useful here. Not shown is the above-mentioned feed preheater which reclaims part of the heat of solution of evaporated ammonia in weak-aqua absorbent. It is, most likely, placed on stream F_1 (Fig. 5), ahead of the intermediate condenser.

The dotted lines in Fig. 5 indicate a way for extending the utility of a falling-temperature heat source. Such an extension may also be had by splitting the single reboiler of Fig. 5 into multiple units, located at separate, higher elevations in the column. The higher ammonia content of the upper trays lower the inlet temperature to the top reboiler.

The three cases of this study were all based on a fractionator bottom pressure of 1413 kPa (205 psia). The reboiler design data for these cases are:

Lastly, in designs where the preheated temperatures of streams F_1 and F_2 in Fig. 5 differ appreciably, they are fed to separate entry points in the fractionator to obtain the efficiency increase reported by Holldorff (6).

HEAT SOURCES

The lack of specificity on design details and payout times herein is due to the great variety of heat sources (4) that may be used. Their primary qualification is that they be available at a temperature high enough to 'drive" the reboilers and preheaters of the distillation unit, and offered at a value well below the unit cost of purchased fuel. If the quantity made available is not adequate to provide the total refrigeration demand, any economic benefit derived from the use of the AAR technique will, of course, be diluted.

The nature of the heat source(s) employed has only obvious constraints (e.g., toxicity, corrosiveness, interruptibility). Most desirable is a source within the process itself, or its auxiliary/ancillary operations. A constant-temperature heat source (such as exhaust steam) is obviously the easiest to use. As mentioned, multiple series reboilers can be used to extend the utility of a falling-temperature heat source. The process-side inlet and outlet temperatures for a single (once-through) reboiler and for three reboilers in series are given in Table 4, also the composition of the corresponding fractionator bottoms. These data are useful in integrating the AAR unit and the heat energy input.

Case	A	B	C
Reboiler duty			
GJ/h	274.5	110.4	53.85
10^6 Btu/h	260.2	104.6	51.04
Temperature °C/°F			
Column bottom	149/300	149/300	132/270
Reboiler inlet			
Single	132/270	116/241	85/185
Multiple	131/267	111/233	79/174

The fractionator bottom temperatures in these designs were arbitrarily chosen in the absence of a specific temperature history of the prime heat source. Maximizing this temperature to the extent possible reduces the requisite flow of weak-aqua and significantly increases the thermal efficiency of the AAR unit.

GENERALIZED AAR EFFICIENCIES

Fig. 2 is plotted from detailed calculations setting the interstage flows, temperatures, and pressures in the refrigerant compression cascades of Case A, based on a single fixed load at -162°C (-260°F). The AAR unit of this case is designed to accept this load, plus the actual work of compression to permit condensation of ethylene refrigerant vapor at -34.4°C (-30°F). It consumes 274.5 GJ/h (260.2 x 10^6 Btu/h) of heat energy in providing this refrigeration. For a given intermediate refrigeration temperature in the methane or ethylene cascade, these calculated flows and their enthalpies define an equivalent single-load refrigeration duty which requires the same AAR output and heat consumption. Dividing this fixed AAR heat input by the equivalent duty gives its specific heat consumption for providing refrigeration at this temperature level. Fig. 6 is a plot of these data. It shows, for example, that transferring the basic load of Case A, 40.56 GJ/h (38.44 x 10^6 Btu/h), at the higher temperature of -120°C (-184°F) (rather than 162°C/-262°F) would require 4.0 x 40.56 = 162 GJ/h (154 MBtu/h) of heat input to the AAR unit, or 59% of that for methane liquefaction.

LITERATURE CITED

1. Bogart, Marcel J. P., "Ammonia Absorption Refrigeration in Industrial Processes." Gulf Publishing Co., Houston (1981).
2. Ibid., p. 103.
3. Ibid., p. 195.
4. Ibid., Chapter 16.
5. Bogart, Marcel J. P., "Pitfalls in Ammonia Absorption Refrigeration." Presented at the Second World Congress of Chemical Engineering, Montreal, Oct. 6, 1981; Plant/Operations Progress, 1, 147 (July, 1982); International J. Refrigeration, 5, 203 (July, 1982).
6. Holldorff, Gunther. "Revisions Up Absorption Refrigeration Efficiency," Hydrocarbon Processing, p. 149 (July, 1979).
7. Picciotti, Marcello, "Optimize C_2H_4 Plant Refrigeration." Hydrocarbon Processing, p. 157 (May, 1979).

ACKNOWLEDGMENT

The assistance of Fluor Engineers, Inc. in the preparation of this manuscript is gratefully acknowledged.

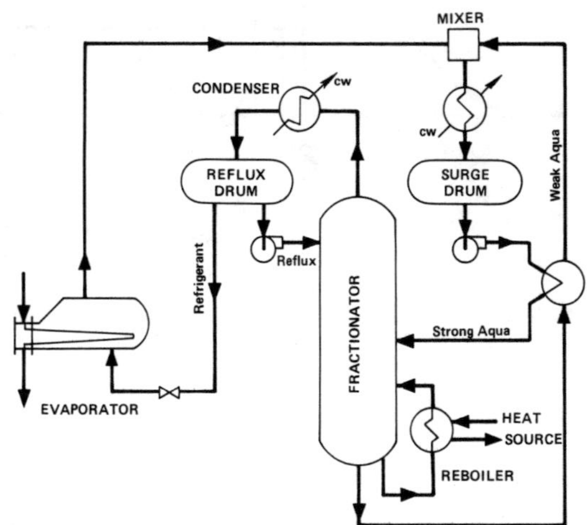

Figure 1. Basic ammonia absorption refrigeration unit.

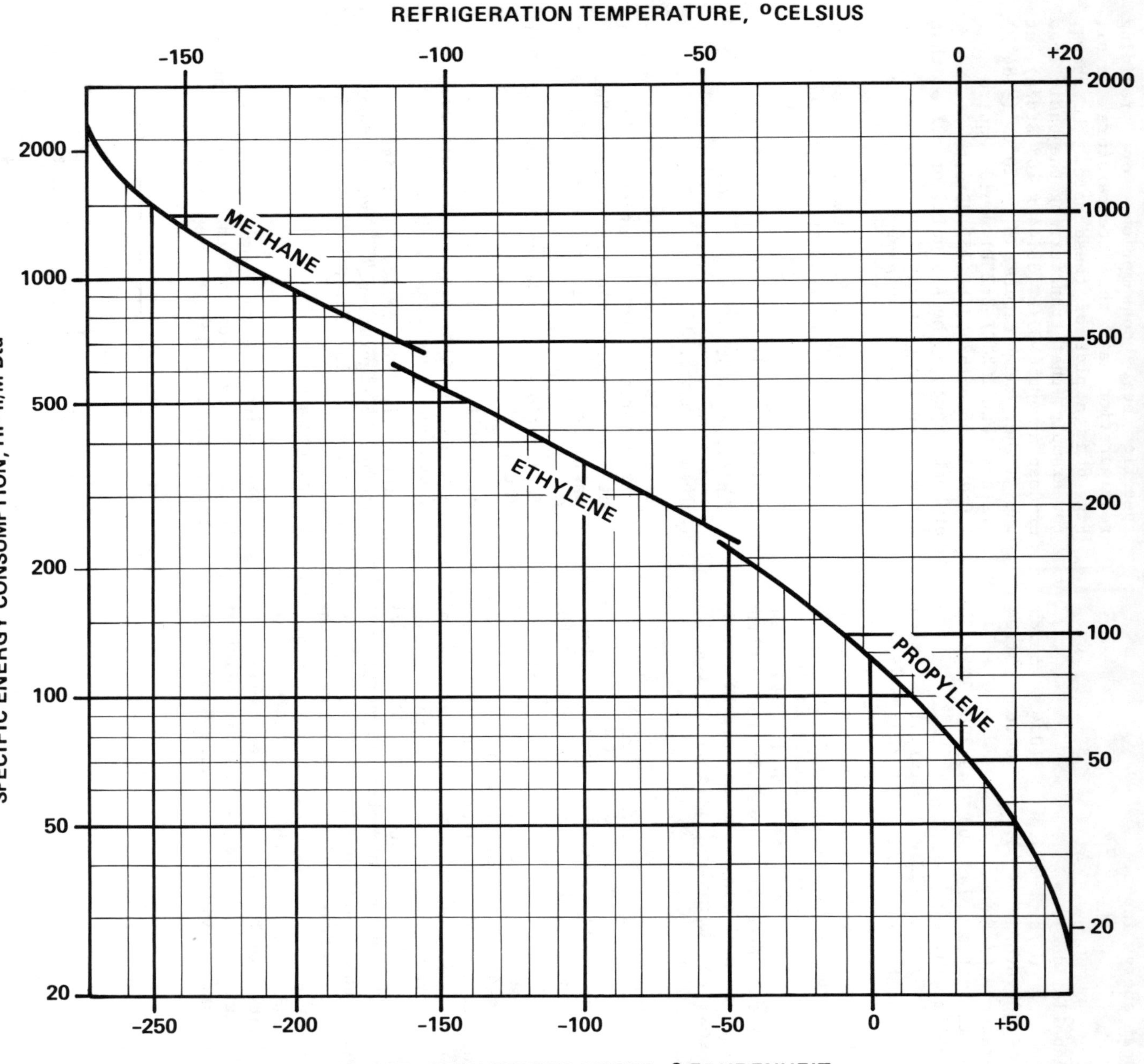

Figure 2. Specific work energy for mechanical refrigeration.

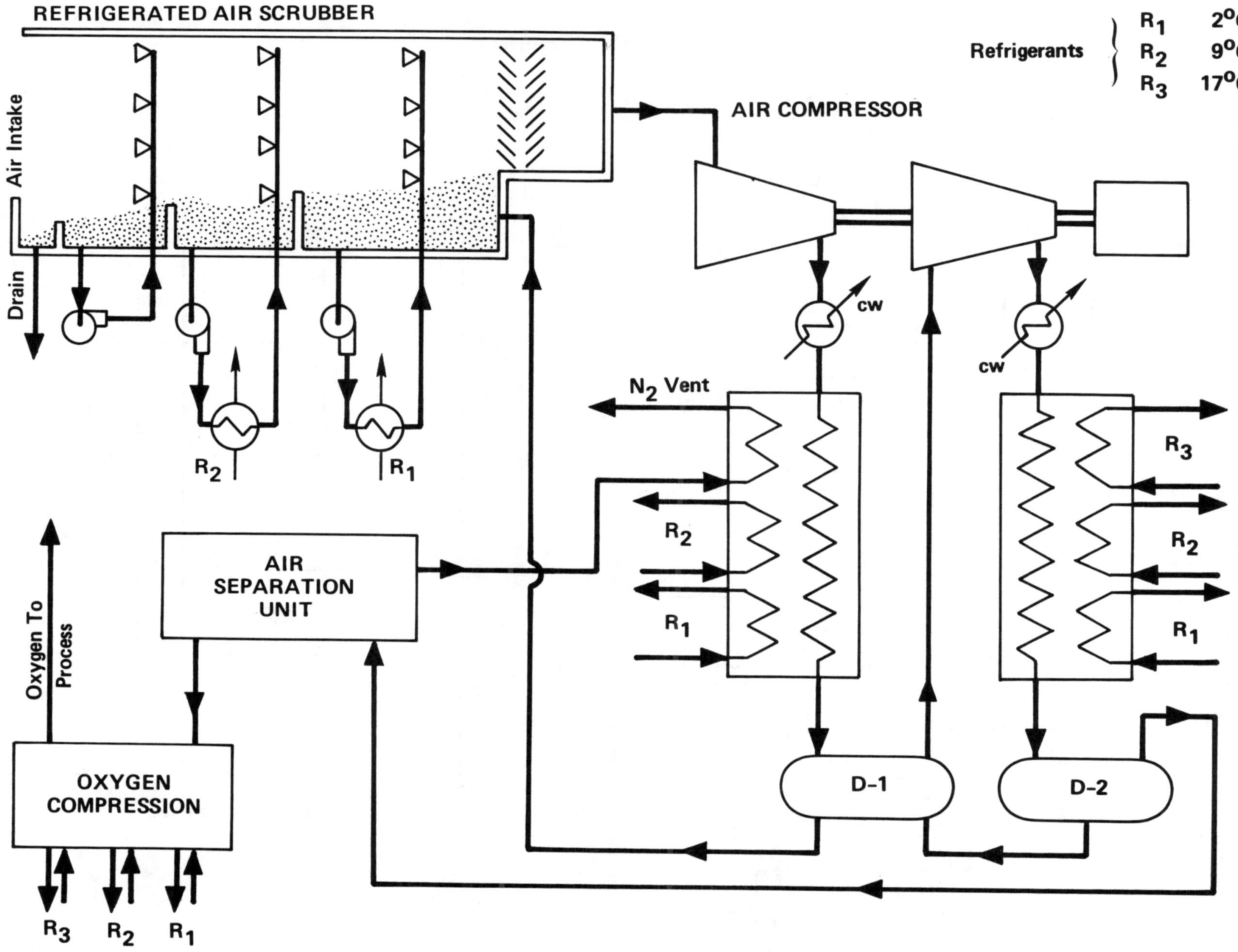

Figure 3. Oxygen plant—air compression section.

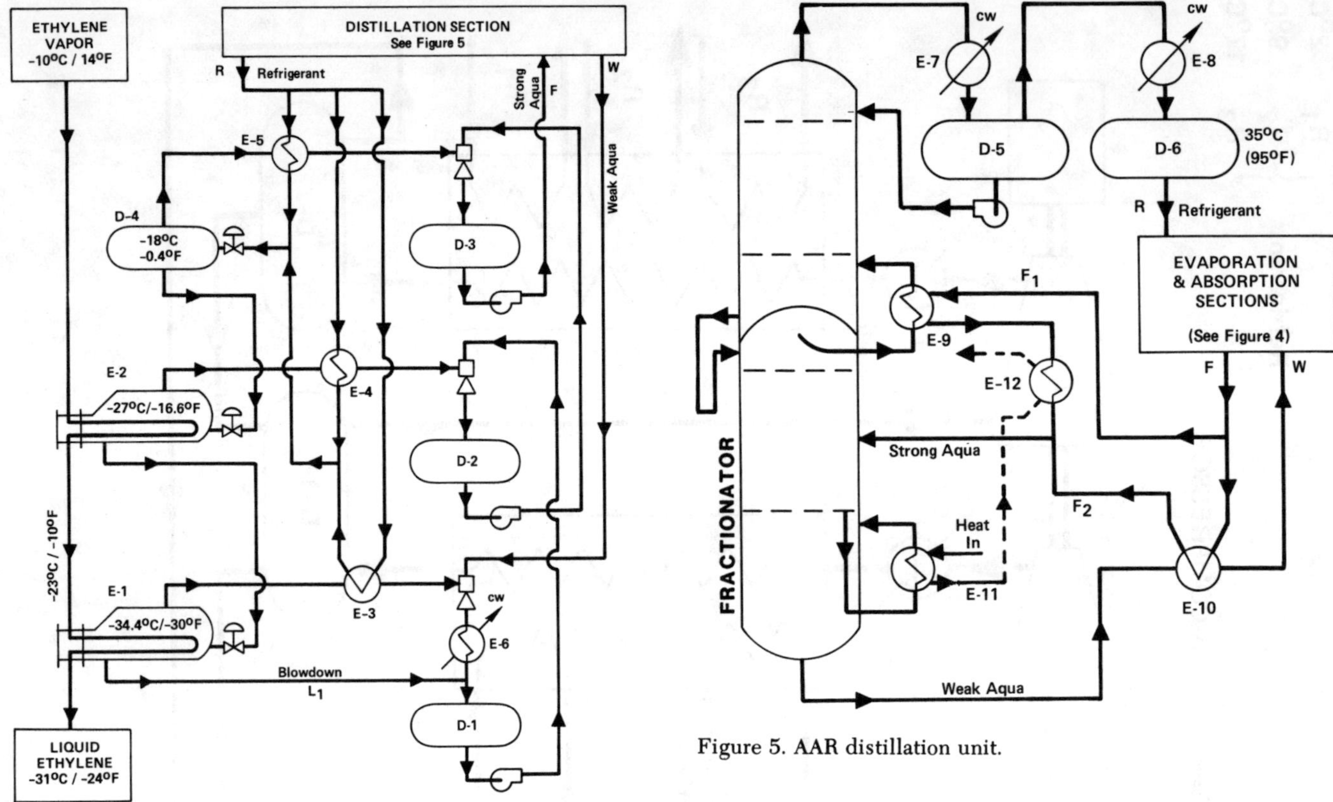

Figure 4. AAR evaporation/absorption cascade.

Figure 5. AAR distillation unit.

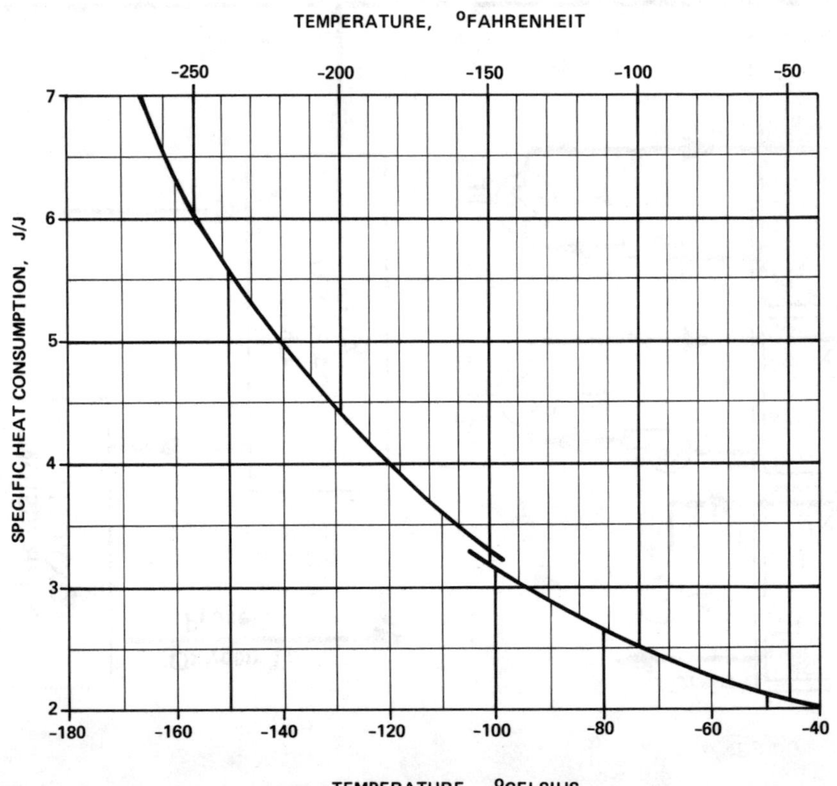

Figure 6. Specific heat energy for cryogenic high-stage AAR unit.

10K LOW POWER CLOSED CYCLE CRYOCOOLER

F. W. PIRTLE

CTI-CRYOGENICS
266 Second Ave.
Waltham, MA 02254

Advances in superconducting electronics have made cryocoolers attractive for many new applications (1). One important application is superconducting magnetometers which can achieve sensitivities on the order of 10^{-12} gauss/ $\sqrt{Hz}$. Successful application of this technology requires the use of a cryogenic refrigeration system that can maintain the low temperature environment while not interfering with the magnetic field measurements. The cryogenic system power requirement, weight, and size must also be compatible with the system requirements of the magnetometer.

A study has been conducted to develop concepts for small, low power, very low magnetic signature, light weight, closed cycle cryocoolers. The design requirements applied to the study are:

- o a net capacity of 50 milliwatts at 10 Kelvin,
- o less than 250 watts of input electrical power,
- o less than 8.5K with no net thermal load,
- o a temperature stability at the load of ± 0.01K,
- o minimum weight, volume, and magnetic signature.

DISCUSSION

Summary

The Stirling cycle was used to achieve the input power requirement because of its very high thermodynamic efficiency. Nonmetallic construction has been used in the cold finger of the Stirling cycle cooler to minimize magnetic interference. Also, the relatively low thermal conductivity of the nonmetallic material reduces the internal conduction loss. The use of paramagnetic materials minimizes magnetic signature at the application interface. The ferromagnetic metals necessarily used in the motor and compressor components can be magnetically shielded to minimize their effect.

Thermodynamic Modeling

The thermodynamic model was developed around the Stirling cycle by modifying existing CTI-CRYOGENICS computer models. No material restrictions were applied at the outset. Regenerator configurations included the annular wall regenerator using wire mesh or spherical particles and the gap regenerator recently under investigation by Zimmerman (2).

Annular Wall Regenerator. As described in the literature (3), an annular wall regenerator consists of conventional wire mesh or spherical particles enclosed in a hollow annular cylinder wall. Application to Stirling systems is illustrated in Figure 1 for a six stage system.

The annular regenerator offers an advantage over the standard regenerator-in-displacer because it is a nonmoving metallic regenerator. A non metallic displacer can be used to avoid the magnetic interference caused by metallic movement in the Earth's magnetic field. A seal is applied at the top end of each displacer to prevent helium leakage through the gap between the cylinder and displacer. Only the first stage seal is illustrated.

Gap Regenerator. The gap regenerator is illustrated in Figure 2, again for a six stage system. Regeneration is accomplished by thermal penetration into and out of the cylinder and displacer walls as the helium flows through the gap between them. Gaps are on the order of 0.010 cm radially for a displacer diameter of 2 cm. The thermal penetration is in the form of a damped temperature wave propagating into the walls as the helium is cooled and out when the helium is heated. The basic model is given in Jakob (4). The temperature wave amplitude, θ, is given by:

$$\theta = \theta_{o,a} e^{-mx} \tag{1}$$

where $m = (\pi f/\alpha)^{1/2}$ (2)

The thermal penetration thickness of the regenerator wall can be represented by a thickness $\lambda = (\alpha/\pi f)^{1/2}$ where temperature is constant over λ. In this way the same amount of heat is stored in the λ thickness as in an infinite wall over one cycle. In this study the total wall thickness for structural reasons was at least four times that needed for thermal regeneration. The penetration depth λ increases at lower temperature, particularly below 20K because thermal diffusivity increases.

Gross Refrigeration. The gross refrigeration is given by:

$$Q_g = \oint PdV \tag{3}$$

Losses. The losses applied to the gross refrigeration are thermal conduction, shuttle or motional, pumping (not a loss in gap regeneration), regenerative friction, and regenerator. These losses have been described in the literature (5, 6).

Net Refrigeration. The net refrigeration is the gross refrigeration less the sum of the losses.

Thermal Damping. Thermal oscillation at the load interface due to refrigerator cycling was damped to within 0.01K using a block of lead, a thermal damping block, placed between the load and the refrigerator cold tip. In order to model thermal oscillation, the damped temperature wave model described earlier and given in Jakob (4) was used. The temperature cycle amplitude of the inside end of the cold finger based on the helium gas temperature amplitude, $\theta_{e,a}$, is:

$$\theta_{o,a} = \theta_{e,a} n \tag{4}$$

where:

$$n = (1/(1 + 2m/b + 2m^2/b^2))^{1/2} \tag{5}$$

and b = k/h. The surface heat transfer coefficient, h, was evaluated using the General Electric correlation (7).

Then the temperature amplitude at λ measured from the helium gas interface in the damping block is:

$$\theta = \theta_{e,a} n e^{-mx} \tag{6}$$

In addition to damping thermal oscillations to ± 0.01K, the damping block must minimize the temperature drop between the electronics and helium gas. Therefore, the damping block must have a high thermal conductivity and low thermal diffusivity. Lead was chosen as the best material to satisfy these requirements. Also, lead is diamagnetic above the superconducting transition temperature of 7.2K; therefore, it is magnetically acceptable. With an empirically based maximum gas temperature excursion at the cold tip of ± 2K, the dimensions of the lead damping block were adjusted to provide sufficient thermal damping with an acceptable temperature drop of 0.1K.

Thermodynamic Trends

Several parameter variations were performed on the annular wall regenerator with the object of reducing the input power below the study goal. Using a 50% motor efficiency and a power transmission loss factor to include mechanical losses, the lowest predicted input power was 1200 watts. The major thermodynamic loss was thermal conduction primarily in the regenerator matrix and secondarily in the stainless steel walls. Plastic or glass walls will reduce the wall loss somewhat; however, the input power was not competitive with gap geometries. Therefore, the annular wall geometry was discarded in favor of the gap geometry.

Trends from a thermodynamic study of the gap geometry are shown in Figures 3, 4, and 5 using MACOR™, a machinable glass ceramic made by Corning Glass Works Inc. Figure 3 shows the trend of input power as a function of the number of stages. The motor efficiency and mechanical loss factor applied to the annular wall system have been applied to all gap results also. Figure 3 shows a decrease in input power with an increase in the number of stages, but with diminishing returns.

The trend of the thickness, $\lambda = (\alpha/\pi f)^{1/2}$, versus temperature is shown in Figure 4. The specific heat decreases faster than thermal conductivity with decreasing temperature; therefore, λ increases with decreasing temperature. The result is that the colder stages must have thicker walls in order to accomplish the thermal regeneration. Structural requirements dictated that the walls be thicker than thermodynamically required. Thermal conduction loss in the thicker lower temperature walls is partially offset by the lower thermal conductivity.

The trend of the λ depth versus frequency is shown in Figure 5. This trend indicates that higher operating speeds have the advantage that thinner regenerator walls can be used to accomplish the required regeneration.

Concepts

Several material combinations were examined for thermal performance using the gap regenerator. Examples include but were not limited to displacer/ cylinder combinations of nylon/stainless steel, quartz/quartz, and nylon/MACOR™.

The three most promising concepts thermodynamically are high speed and low speed concepts using MACOR™ and a nylon concept. An integral drive system (displacer and compressor driven by same crankshaft) was used. A split system (gas or fluid driven displacer) has advantages of potentially lower magnetic fields at the electronics and greater packaging flexibility; however, the integral system has more consistent hardware performance and a lower power requirement. Table 1 gives the specifications of the concepts. Figures 6, 7, 8 are concept drawings.

Construction of the three concepts is similar with the displacer and cylinder walls constructed of paramagnetic nonmetallic materials. A thin metallic coating must be applied to the outside of the cylinder wall to prevent helium permeation through the wall when any non metal is used. Therefore, except for potential vibration of radiation shields, vacuum shell, lead damping blocks and metallic coating to eliminate helium permeation, there are no moving metal parts in the vicinity of the electronics.

Four stages of refrigeration are used. Radiation shields are attached to the first stage (180K) and second stage (80K). The space inside the vacuum vessel is filled with aluminized mylar superinsulation. The compressors shown with the concepts are representative only.

Nylon Concept. Nylon appears attractive thermodynamically because its low thermal conductivity produces low conduction losses and yet allows acceptable regeneration. A nylon concept is shown in Figure 6. Unfortunately, the low rigidity of nylon will result in a larger vibrational amplitude at the cold space due to stretching because of both pressure cycling and laterally induced vibration. The elastic modulus of nylon is less than 1/10 that of MACOR™.

MACOR™ Concept. A concept embodying the use of MACOR™ cylinder walls and displacer and a relatively slow operating speed of 100 RPM is shown in Figure 7. The longer length of the concept corresponding to the slow speed has the potential advantage of greater distance between motor magnetic fields and the cold electronics space; however, proper magnetic shielding around the motor should greatly reduce this advantage over the smaller higher speed MACOR^{TM} concept.

Small MACOR Concept. Figure 8 shows a MACOR^{TM} concept with an operating speed of 2000 cycles per minute. This relatively high speed makes this concept appreciably smaller than the other two concepts. The required refrigeration can be produced with a smaller PV diagram (smaller expansion volume) at higher speed. While losses do not necessarily change directly with speed, the losses can be satisfied with a smaller system at the higher speed. Also, the thermodynamic regeneration thickness is smaller with higher speed as illustrated in Figure 5. These factors generally provide a smaller and lighter weight system at higher speeds.

CONCLUSIONS

The conclusions from the study are listed as follows:

- o The small MACOR^{TM} concept is the most attractive concept for furthur development.
- o Cryocoolers using a series of gaps as regenerators appear attractive for low capacity, low input power applications.
- o Nonmetallic materials can be used for regenerators.
- o Achievement of 50 milliwatts of refrigeration at 10K with less than 250 watts of input power is possible.
- o Achievement of $\pm$ 0.01K oscillation at the cold space is achievable with proper thermal damping.
- o Paramagnetic materials can be used in cryocooler construction in order to minimize magnetic signature.

ACKNOWLEDGEMENT

This study was supported by the Office of Naval Research under Contract N00014-80-C-0465.

LITERATURE CITED

1. J.E. Zimmerman, D.B. Sullivan, and S.E. McCarthy Eds.: Refrigeration for Cryogenic Sensors and Electronic Systems, NBS Special Publication 607, Supt. of Documents, U.S. Gov. Printing Office, Washington D.C. 20434 (1981).

2. J.E. Zimmerman, and T. Flynn, Eds.: Applications of Closed-Cycle Cryocoolers to Small Superconducting Devices, NBS Special Publication SP-508, Supt. of Documents, U.S. Gov. Printing Office, Washington D.C. 20434 (1978).

3. H.L. Kroebig, 20K Cooling System for High Performance FLIR, Report No. AFWAL-TR-81-3014, Contract F33615-77-C-2049, Flight Dynamics Laboratory, Air Force Wright Aeronautical Laboratories, Wright Patterson Air Force Base, Ohio 45433 (1981).

4. M. Jakob, Heat Transfer, Vol. 1, Wiley (1962).

5. W.R. Martini, Stirling Engine Design Manual, Report No. NASA CR-135382 (NTIS N78-23999), Contract NSG-3152, U.S.D.O.E., Washington, D.C. 20545 (1978).

6. G. Walker, Stirling Engines, Clarendon Press, Oxford (1980).

7. R.E. Norris, Ed., Heat Transfer Data Book, General Electric Co., Schenectady, N.Y. (1977).

NOTATION

Symbol		Definition	Units
b	=	k/h	cm^{-1}
c_p	=	specific heat	j/g-K
f	=	frequency	Hz
h	=	surface heat transfer coefficient	$j/sec\text{-}cm^2\text{-}K$
k	=	thermal conductivity	j/sec-cm-K
m	=	logarithamic decrement	cm^{-1}
p	=	pressure	atm
Q_g	=	gross refrigeration	watts
Q_n	=	net refrigeration	watts
V	=	volume	cm^3
x	=	wall thickness coordinate	cm
α	=	thermal diffusivity $= k/\rho c_p$	cm^2/sec
λ	=	thermal penetration thickness	cm
ρ	=	density	g/cm^3
θ	=	temperature wave amplitude	K

Subscript

a	=	denotes condition at a particular time
e	=	denotes condition in gas in contact with wall
o	=	free surface, $x = 0$

Table 1. Concept Specifications

Specification/Concept	Nylon	MACOR	Small MACOR
Input Power, Watts	142	217	166
Stage Capacity, Watts*	0.277/0.062/ 0.111/0.050	0.316/0.072/ 0.045/0.050	0.164/0.138/ 0.123/0.050
Speed, RPM	60	100	2000
Wall Thickness, cm*	0.16/0.16/ 0.25/0.40	0.16/0.16/ 0.25/0.40	0.08/0.08/ 0.08/0.13
Stroke, cm	0.70	0.95	0.25

Specifications common to all three concepts:

Stage Temperature, K*	180/80/30/10
High Pressure, atm	7.1
Low Pressure, atm	2.8

*Stage specifications are from warm stage to cold stage.

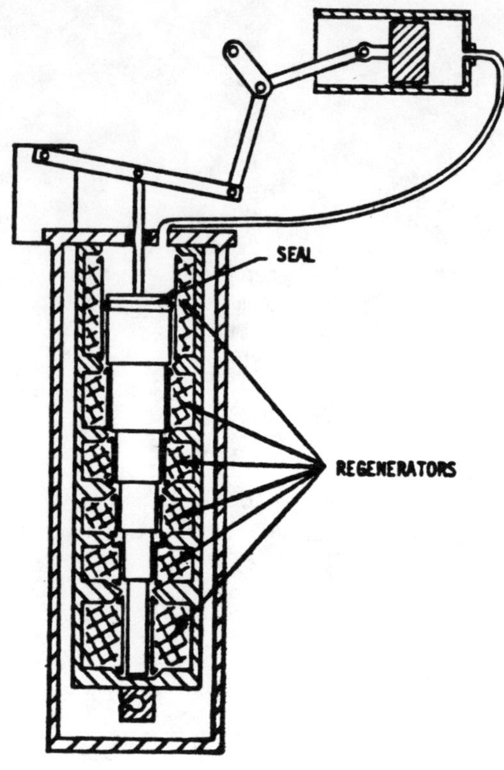

Figure 1. Annular wall regenerator.

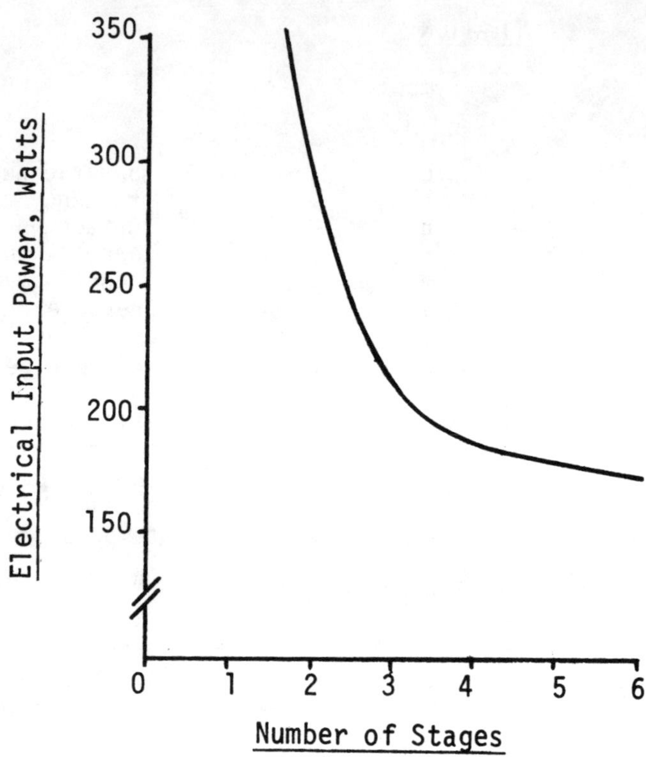

Figure 3. Minimum input power versus number of stages—MACOR™ concepts.

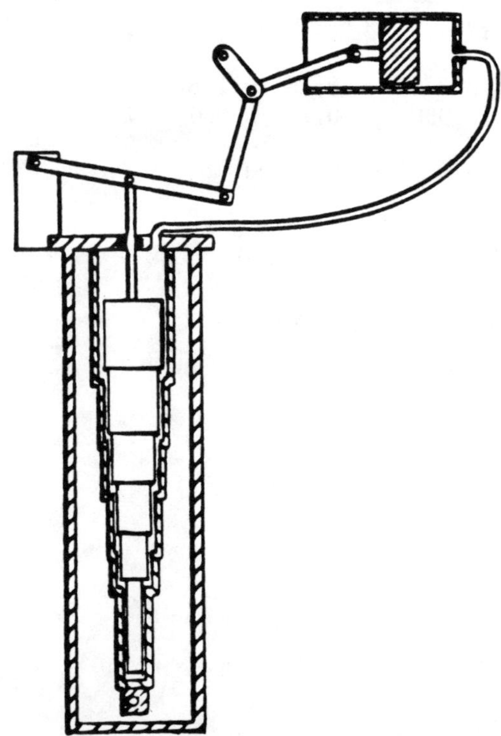

Figure 2. Gap regenerator.

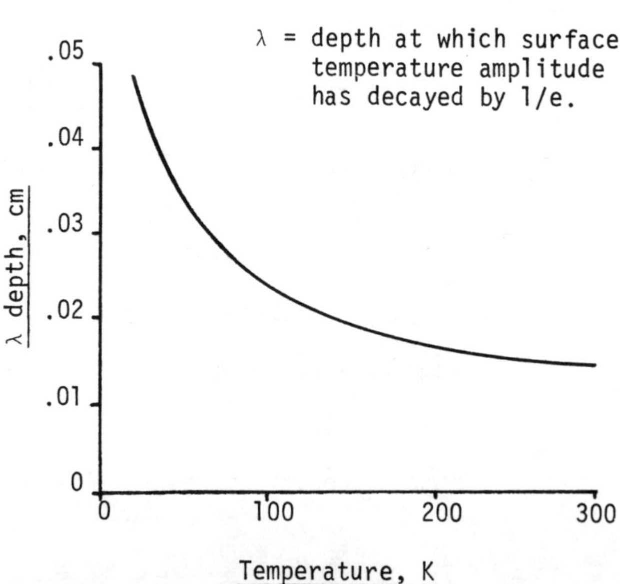

Figure 4. Thermal depth, λ, at 2000 RPM versus temperature for MACOR™.

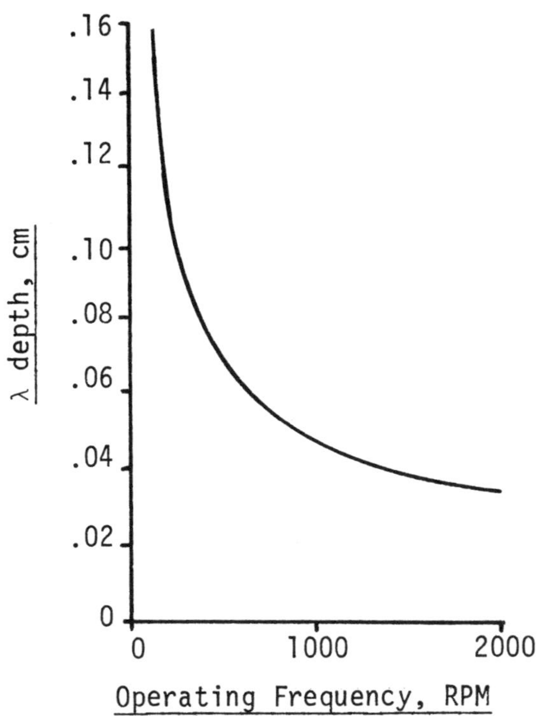

Figure 5. Thermal depth, λ, at 10K versus cycle frequency for small MACOR™ configurations.

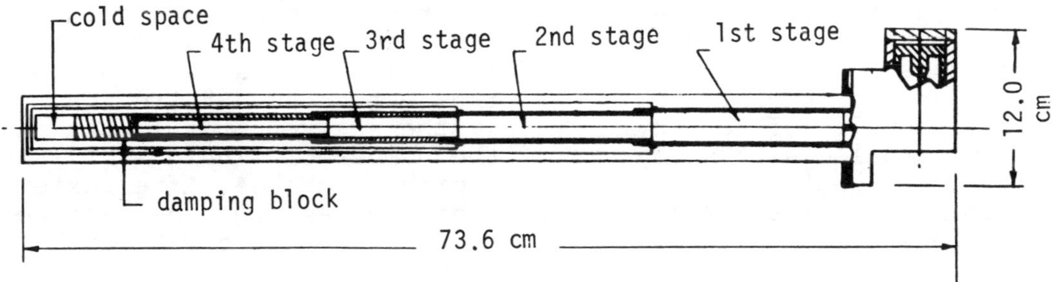

Figure 6. Nylon concept.

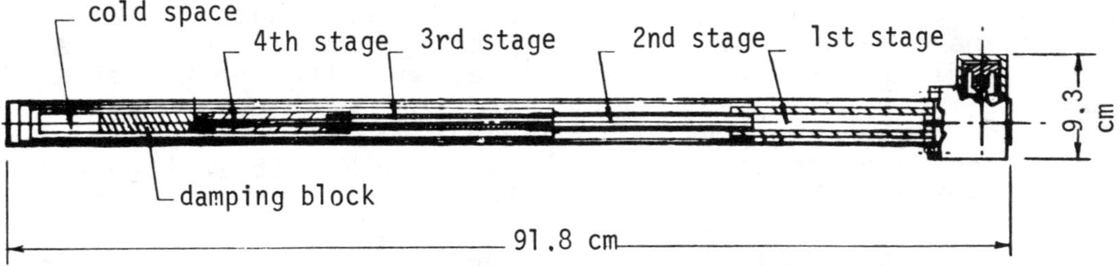

Figure 7. MACOR™ concept.

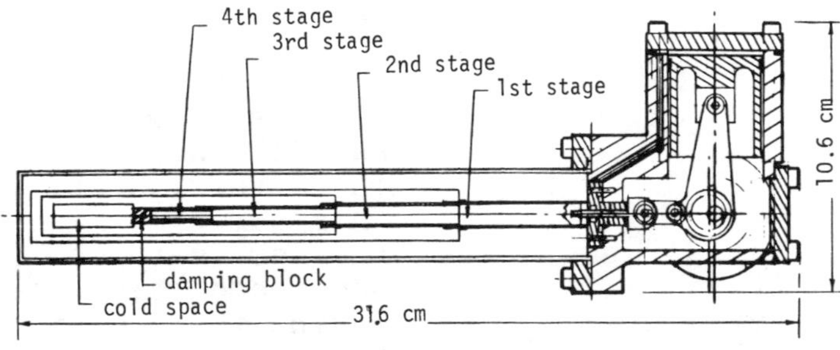

Figure 8. Small MACOR™ concept.

CRYOGENIC TECHNIQUES AND REFRIGERATION FOR A LARGE-SCALE SUPERCONDUCTING GENERATOR

A. STEPHAN
LINDE AG, Munich

Superconducting generators are world-wide under development. In the Federal Republic of Germany, the KRAFTWERK UNION (KWU) pursues an extensive program for the development of superconducting AC generators. In this program, the LINDE AG, Munich, also takes part and is responsible for the cryo-refrigerator.

The LINDE AG has designed a refrigeration system which presents a) an overall concept (process and design) for the refrigeration supply of the probably biggest commercial S.C. generator that is planned to be realized with a capacity of 1,000 MVA and b) the refrigeration supply for the biggest testing rotor being under construction. This refrigeration system has to fulfill requirements which, in some specific aspects, go far beyond those which have to be met by the refrigeration systems built up to now. Examples: An uninterrupted LHe supply for more than 30,000 hours has to be guaranteed, which requires special compressors, valves, etc.; the process has to be controlled in such a way that the refrigeration system is adapted automatically to all different states of operation (cool-down, stand-by, steady-state, etc.); the whole system has to be integrated into the structure of a commercial power station.

The overall design of the system dimensioned for an LHe capacity of 5 g/s (150 l/h) is described as well as its thermodynamic design and the provisions that have been necessary with regard to selection, construction and manufacture of the components in order to reach high reliability of operation. A description of important components such as compressor, oil separation system, cold box with turbines, LHe storage and distribution and computer process control is also given.

Part of the world-wide activities in the development of superconducting generators (1) is the development program of KWU/SIEMENS in the Federal Republic of Germany. In 1973, a conceptual design study was started, investigating the technical feasibility of superconducting generators up to unit ratings of 3,000 MVA (2). To this, LINDE AG contributed by carrying out basic research with regard to the refrigeration supply of superconducting generators and showing solutions by means of examples. Further theoretical, constructive and experimental in-depth investigations of the problems as presented by the conceptual design study were carried out; the results of these investigations as well as a favourable estimate of the economical merits of superconducting generators encouraged KWU/SIEMENS to elaborate the present development program. Its aim is the manufacture of a superconducting 1,000 MVA generator by the end of the eighties.

According to Figure 1 the development of conventional high-capability generators (1 to 3) is continued by the development program for superconducting generators (4 to 6). The manufacture of a full-scale superconducting test rotor "B", reduced only by the active rotor length and the number of windings will be finished in 1984. After extensive separate testing, the superconducting test rotor "B" will first be operated as a superconducting experimental generator (4 in Figure 1) with different stator winding designs and, after that, as a superconducting 120/400 MVA prototype generator (5). A further step will then be the assembly and test of a superconducting generator of the commercial size of 1,000 MVA (6 in Figure 1) with the aid of a new superconducting rotor "C". It is planned to operate this generator after thorough factory testing in a power plant at the beginning of the nineties.

Continuing the close and successful cooperation with KWU/SIEMENS, LINDE AG is developing a refrigeration system (4 a in Figure 1) which will supply the necessary cooling for the superconducting field windings during the aforementioned phases

The program of developing superconducting generators is funded by the German Federal Ministry of Research and Development (BMFT).

of development of the superconducting generator. This refrigeration system is described in the following.

REFRIGERATION SYSTEM REQUIREMENTS

For cooling of the test rotor "B" (phases 4 and 5 in Figure 1) as well as of the 1,000 MVA rotor "C" (phase 6), a cooling stream of 5 g/s of liquid helium is required. The helium is fed to the rotor with a pressure of 1.1 bar and a temperature of 4.3 K. It returns with a pressure of 1.05 bar and a temperature of about 250 K.

The cool-down period of the test rotor "B" is to be less than three days and that of the 1,000 MVA rotor "C" less than six days. The masses that have to be cooled down to 4.5 K correspond to 12.5 and 25 metric tons, respectively. Between the reference temperature of the rotor and the cooling gas stream a temperature difference of 30 K is permitted during cool-down. Liquid nitrogen precooling can be used during cool-down.

Operation of the refrigeration system is to be fully automatic. It is of particular importance that the refrigeration system adjusts automatically to both transient and steady-state operations.

The reliability of the refrigeration system has finally to ensure an uninterrupted LHe supply to the generator for more than 30,000 hours, as required in power plant operation.

PROCESS DESCRIPTION

Based on our preliminary works of the years 1975 and 1976, a conceptual design has been chosen for the refrigeration system that is characterized by the well-known reliability of its components. At the same time, it also takes into consideration the need for maintenance works to be carried out as well as the possibility of failures of technical systems that can never be completely excluded by utilizing a LHe dewar and providing several redundant units. Figure 2 shows a block diagram of the system and its main components.

The LHe for the superconducting test rotors as well as the superconducting generator (5) is supplied from the LHe dewar (3). From there the LHe is fed via a small vessel, the so-called cryoterminal (4), to the helium transfer equipment (HTE, 5 a) of the generator. The gaseous helium leaving the generator is fed to the helium liquefier (2) and from there it is reliquefied into the LHe dewar (3).

In case of a temporary shut-down of the liquefier due to maintenance works the LHe supply of the generator can be maintained for more than 24 hours by means of the LHe dewar reservoir. This period is thought to be sufficient for the plant to reach steady-state operation again after an interruption. The helium return during a shut-down of the refrigeration system can be compressed by one of the two compressors of the liquefier (2) into the gas storage tank (1). After the plant has been started again, it can cover not only the coolant requirement of the superconducting generator, but also refill the LHe dewar (3) by means of a temporary increase in production. For this purpose, the liquefier has been designed for a total production capacity of 6 g/s LHe (ie 20 % overdesign).

For the liquefaction process itself a Claude process with two precooling stages and isenthalpic expansion into the two-phase region by means of a Joule-Thomson valve has been chosen. This comparatively simple process presents, together with the application of single-stage, oil-cooled screw compressors, considerable advantages with regard to the required reliability of the plant (3).

Figure 3 shows the process diagram of the refrigeration system of the superconducting 1,000 MVA generator. The compressors (1) deliver 185 g/s of helium compressing it from 1.05 bar to 11 bar. Their power requirement amounts to approx. 500 kW. After being cooled down to about 72 K in the heat exchanger (4) the gas stream is divided. One partial stream of 150 g/s is further cooled down in the two turbines (2 and 3) connected in series, until finally reaching a temperature of 11 K at a pressure of 1.3 bar. The second stream is cooled down to about 7 K in the heat exchangers (5 to 8) and expanded in the Joule-Thomson valve (9) from 10 bar to 1.3 bar at 4.5 K, being partially liquefied. In the LHe dewar (10) the liquid and the gaseous phase are separated. The helium gas returns to the cold box, streams through the heat exchangers (8 to 4) and is warmed up together with the gas stream coming from the turbines and then led to the suction side of the compressors (1).

According to the coolant requirement of the superconducting rotor (13) an LHe stream of 5 g/s is withdrawn from the LHe dewar (10) and led to the cryoterminal (11) which has a pressure of 1.1 bar as determined by the rotor design. From the cryoterminal it is led to the helium transfer equipment (HTE) of the rotor. The helium vaporizes in the rotor, leaves it via the HTE with a temperature of about 250 K and streams with a pressure of about 1.05 bar to the suction side of the compressors (1).

During refilling of the LHe dewar, a helium stream of 1 g/s is led from the gas storage tank (14) to the refrigeration system, cooled down, liquefied and collected in the LHE dewar. Once the dewar is filled, the liquefaction capacity of the refrigeration system is reduced again to just covering the coolant requirement of the generator.

DESIGN OF THE REFRIGERATION SYSTEM AND DESCRIPTION OF ITS COMPONENTS

Figure 4 shows the simplified flow diagram of the complete refrigeration system for the generator. The gas storage tank (D 110), the helium liquefier consisting of two compressors (C 101) and a cold box (D 201), the LHe dewar (D 301), and the cryoterminal (D 302) are the main components of the system as previously shown in Figure 2.

For compressing the gas stream the two oil-cooled screw compressors (C 101) with a capacity of 100 g/s of helium each are used. The compressors are driven by a three-phase AC motor each. Their power requirement amounts to 250 kW each with a final pressure of 11 bar. The heat of compression is led off to the cooling water via the gas coolers (E 101) and the oil coolers (E 102).

For many years, this kind of single-stage screw compressors has been used successfully for the compression of air. To achieve those conditions as necessary for the operation with helium, the screw compressors have been equipped with oil-cooled axial face seals at the primary shaft and O-ring seals at the casing.

For smooth operation of a low-temperature process gas of high purity is required. Therefore, special attention has been given to the oil separation system. By means of the four-stage combined gravity and coagulation oil separation system as installed here (D 101 - D 103) the oil content in the helium is reduced to less than 0.1 ppm by weight. The separated oil is fed back to the suction side of the compressors. By means of the oil adsorber (A 101) installed after the two compressors, the oil content is further reduced to a final value of less than 10^{-2} ppm by weight. While the mechanical oil separation (D 101 - D 103) attached to each compressor does not have a saturation limit, the lifetime of the adsorber is limited to about one year. For this reason, a second adsorber is connected in parallel for alternating operation.

In actual power plant operation of the refrigeration system, a third compressor unit is going to be installed as a redundancy.

The gas storage tank (D 110) consists of four vessels, each having a volume of 100 m^3. It is designed for storing all the gas present in the refrigeration system and the rotor at a pressure of 10 bar and ambient temperature.

In the cold box (D 201) the heat exchangers, the turbines, the adsorbers and the cold valves of the liquefier are arranged vertically. The cold box vessel has the geometry of a lying through with a horizontal cover plate. In the middle of this plate a cylindrical dome is installed which serves for taking up the heat exchangers and the adsorbers.

The heat exchangers (E 201, E 202/203, and E 204/205) are of the aluminium plate-fin type. The five process heat exchangers are combined to three units and fixed in the cold box vessel by means of glass-reinforced polyester elements.

To provide against small quantities of air that might enter the closed circuit coming for instance from the oil system of the HTE bearing of the generator, two alternating adsorbers (A 201) are installed in the cold box. The whole high-pressure gas stream passes through one of them at a temperature of 72 K. The running period of one adsorber amounts to 300 hours in case of an assumed air content of 3 ppm by volume. The loaded adsorber is regenerated by means of heating with an electric heater, evacuation and purging with pure helium.

The turbines (X 201 and X 202) of the liquefier are single-stage centripetal turbines and have dynamic radial and axial gas bearings. For starting the turbines and switching them off, a static auxiliary gas bearing is used. The work of the turbines is taken up at a speed of 150,000 and 90,000 1/min respectively by a helium compressor which is situated on the same shaft. The heat from this compressor is transfered via a cooler to the cooling water.

The insulation of the cold parts inside the cold box is accomplished by means of a multilayer superinsulation in high vacuum. The high vacuum is produced and maintained by means of a pump unit attached to the cold box.

Transport of the cold gaseous and liquid helium between the refrigeration system, LHe dewar, cryoterminal, and HTE of the superconducting generator is done by means of vacuum-superinsulated transfer lines. The first mentioned transfer lines are of the rigid type with a single core since it is not being subject to any strains from outside, whereas the connecting lines between the stationary cryoterminal and the HTE on the free end of the generator shaft have to be flexible and equipped with two cores. This is achieved by an arrangement of four concentric flexible tubes. Cryoterminal and flexible line to the HTE form one unit provided with a joint vacuum.

All the vacuum spaces of the transfer lines and the cryoterminal are constantly connected with a high-vacuum pump unit not shown in this figure. The transfer line between LHe dewar and cryoterminal is equipped with a valve for controlling a constant liquid level in the cryoterminal. This line is of special importance for an undisturbed LHe supply of the generator, therefore a second redundant line is installed to prevent any failures.

The LHe dewar (D 301) is provided with a radiation shield which is cooled by waste gas. It is equipped with a redundant device for pressure build-up in case of a standstill of the liquefier.

The helium vaporizing in the rotor passes through torque tube extensions and current leads and leaves the HTE in three partial streams via flexible lines. According to measuring values obtained from the generator, these partial streams can be controlled individually. Before streaming through the electric heater (E 303), the three streams are combined into one. This stream is heated up to ambient temperature in an electric heater (E 303) and then led to the compressors (C 101).

The refrigeration system will be monitored and controlled by means of a microprocessor system provided with about 100 digital inlets and outlets, about 30 analog inlets and about 20 analog outlets. For recording and storing of the process data, a data transfer to a central process computer is planned.

Besides controlling the process parameters during steady-state operation, the microprocessor system also serves as the controller of the refrigeration system during transient operation. This is for instance: cool-down of the rotor (here the flow rate has to be as high as possible, whereas the temperature difference is not allowed to exceed a certain value); LHe supply to the HTE (the level of liquid helium is dependent on the generators turning speed); change-over and regeneration of the N2/O2 adsorbers (A 201).

The microprocessor system will be provided in redundancy in actual power plant operation.

SUMMARY

For LHe cooling of the superconducting field winding of a big test rotor as well as of a superconducting 1,000 MVA generator, a refrigerator system is being developed. This system will be ready for application by 1984. The helium liquefier of the system is designed for a capacity of 6 g/s; it is equipped with oil-cooled screw compressors, turbines with dynamic gas bearings and a J-T expansion valve. In order to maintain the LHe supply during shut-down periods of the liquefier, an LHe dewar is installed between the latter and the superconducting generator. The refrigeration system is designed for continuously supplying liquid helium to the generator during more than 30,000 hours.

LITERATURE CITED

1. D. Lambrecht, "Status of Development of Superconducting AC Generators", 7. International Conference on Magnet Technology Karlsruhe, BRD, März/April 1981

2. Abschlußbericht der KRAFTWERK UNION AG und SIEMENS AG zum Projekt NT 4443 A des Bundesministeriums für Forschung und Technologie (1976), Elektrische Maschinen mit supraleitenden Wicklungen, Band 1: Turbogeneratoren

3. Schlußbericht der LINDE AG zum Projekt ET 4191 A des Bundesministeriums für Forschung und Technologie (1980), Kälteversorgung supraleitender Generatoren in Großkraftwerken

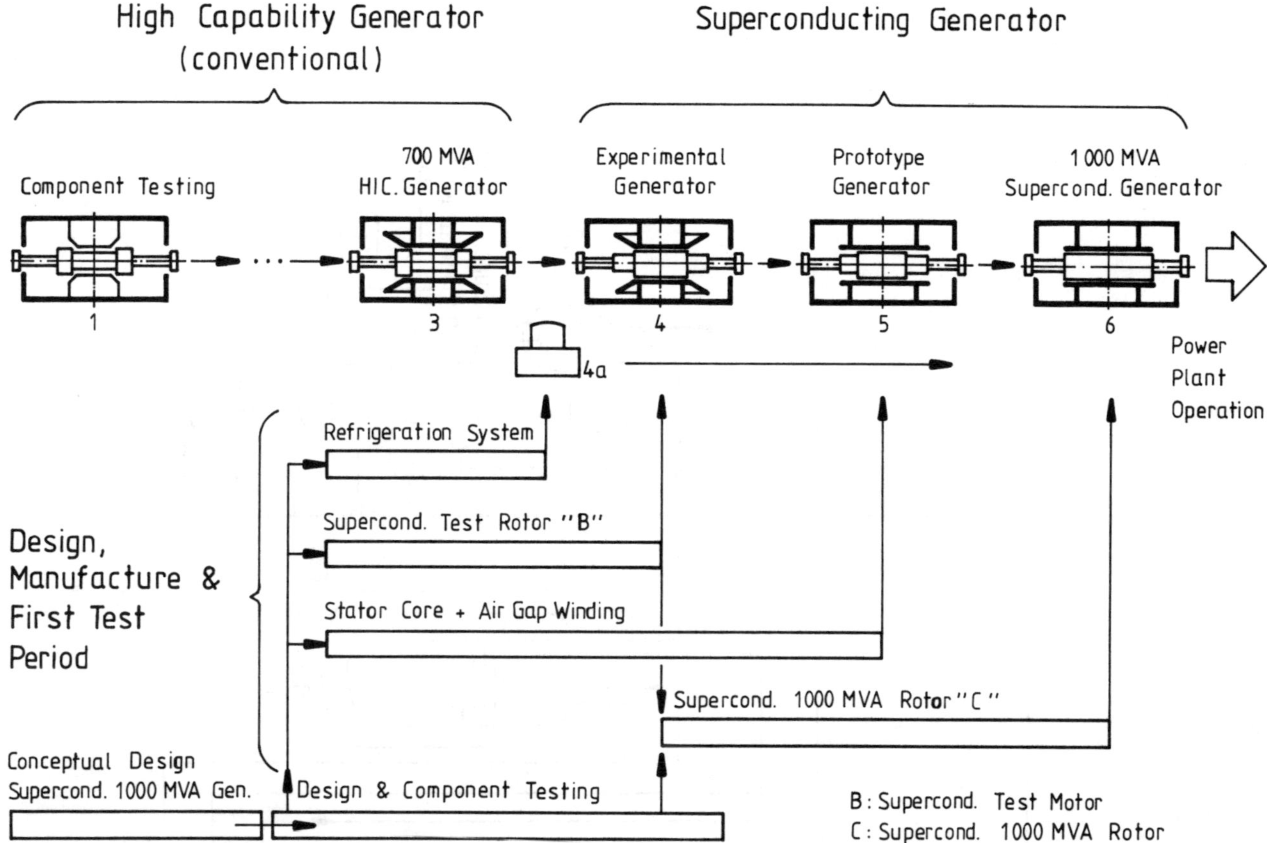

Figure 1. Development program superconducting generator.

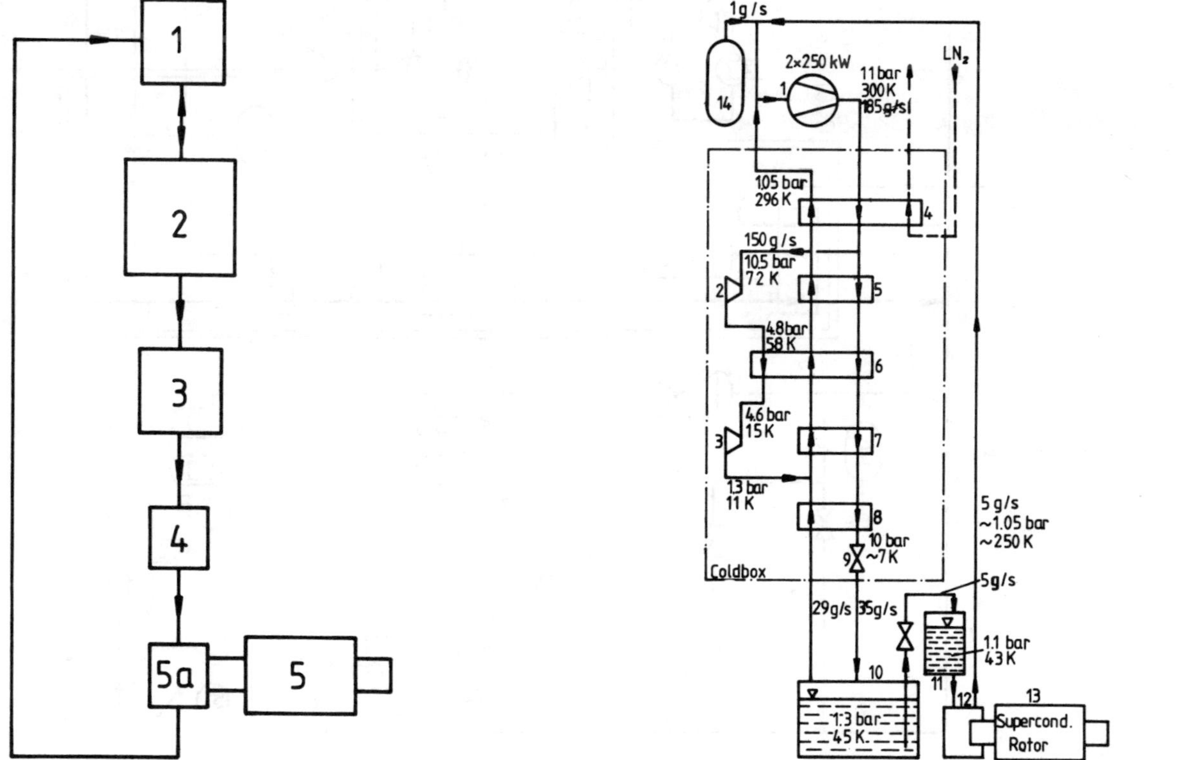

Figure 2. Block diagram of the refrigeration system.

Figure 3. Process diagram of the refrigeration system.

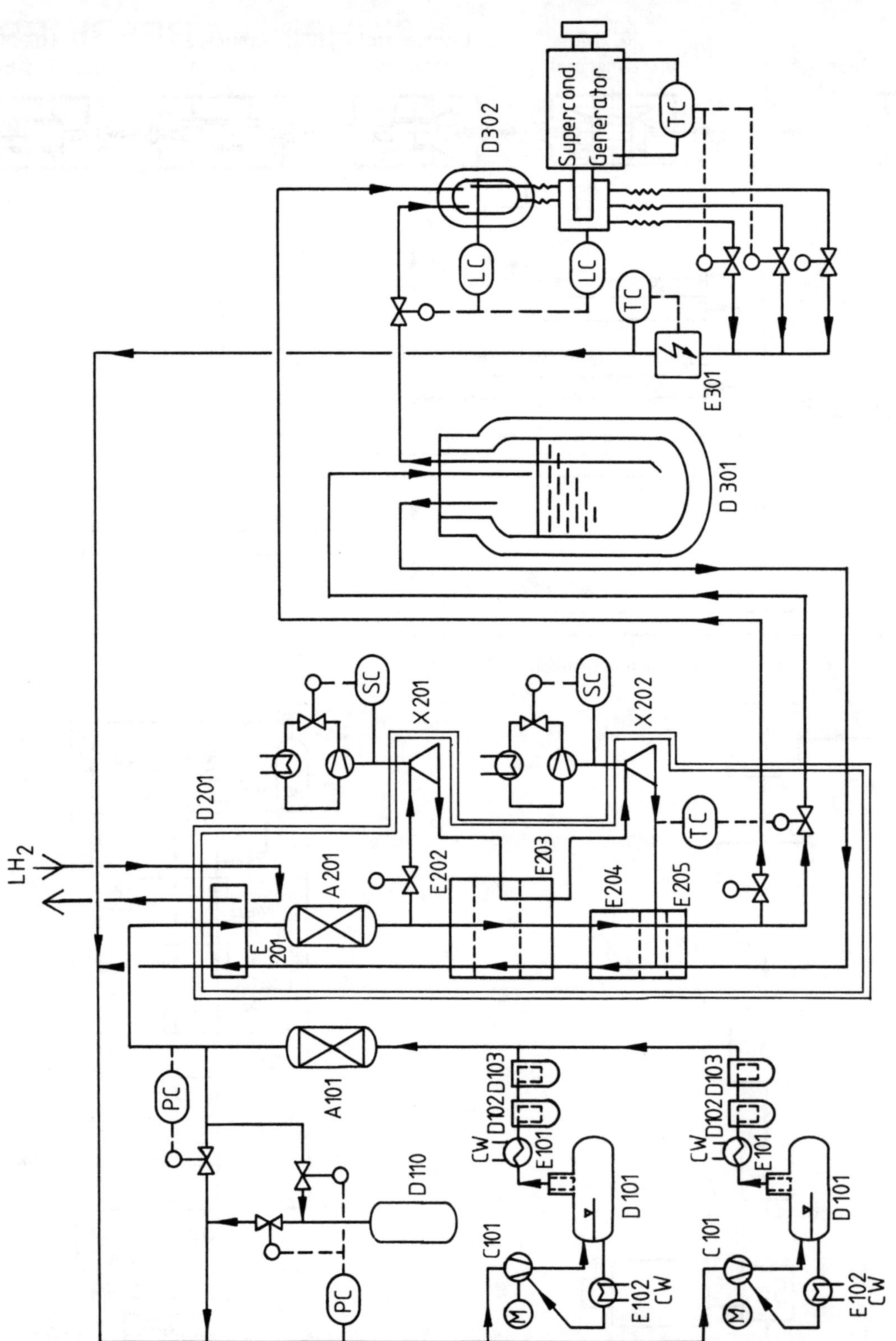

Figure 4. Flow sheet of the refrigeration system for a superconduction 1000 MWA generator.

THERMODYNAMIC EVALUATION OF IMPROVEMENTS TO THE JOULE-THOMSON LOOP OF STEADY FLOW 4.2K HELIUM REFRIGERATORS

P. A. LESSARD
J. M. KAUFMAN
CTI-CRYOGENICS
Kelvin Park
Waltham, MA 02254

Conventional steady flow helium liquefiers and liquid helium temperature refrigerators consist of a high pressure helium stream that is precooled to approximately 10K by either a separate regenerative type refrigerator, a cascade of progressively lower temperature refrigerators, or by portions of the process stream that have been cooled by work extraction. Final expansion of the process stream to the required working pressure is accomplished by throttling the stream through a Joule-Thomson valve after final precooling in a recuperative heat exchanger. This highly irreversible isenthalpic process is thermodynamically wasteful. Three alternative schemes designed to decrease the irreversibility associated with the J-T process are:

- o Expansion engine exhausting in the two phase regime.
- o Expansion engine combined with a saturated vapor (cold gas) compressor.
- o A cold ejector or jet pump.

The purpose of these schemes is to reduce the entropy generated at the cold end by using the high pressure to produce work, which is then either removed or used to advantage in compressing the cold vapor.

BASE CASE-PERFORMANCE OF J-T LOOP

The final expansion stage (J-T loop) of a conventional 4K refrigerator consists of a recuperative (J-T) heat exchanger, an expansion valve, and the refrigeration load. The liquid fraction yield of the J-T valve is a function of the upstream conditions at the valve, which are, in turn, a function of the temperature level, T, and the temperature difference, ΔT, at the top of the J-T heat exchanger. Heat exchanger performance is partly characterized by its effectiveness ε, the ratio of actual heat transferred to maximum potential heat transfer but it is important to note that effectiveness alone does not define heat exchanger performance - two of the three parameters T, ΔT, and ε are required. The analysis of J-T loop performance consists of the following steps:

- o For given refrigeration load conditions, set T and ΔT
- o Use the first law applied to the J-T loop to define flow required
- o Use the second law to check for impossible conditions
- o Calculate heat exchanger effectiveness.

Figure 1 is a schematic of the J-T loop and the results of the performance analysis using, as the figure of merit:

θ =(work required at warm compressor to compress J-T flow)/(refrigeration load)

θ is plotted against ε_{HX} for various T and ΔT. The trends illustrated by this performance map are:

- o Increasing T or ΔT (i.e. lower heat exchanger effectiveness) decreases performance
- o A minimum ε_{HX} of $\sim$ 0.88 is required to produce any refrigeration
- o Above approximately ε_{HX} = 0.96, the performance is relatively flat - that is, the addition of heat exchanger surface does not appreciably affect J-T loop performance.

This performance map is the base case against which the proposed improvements are to be compared.

TWO PHASE EXPANSION ENGINE

If the J-T valve be replaced by an engine of some sort, e.g. piston or turbine, the final expansion takes place more nearly isentropically with a resultant higher liquid yield and greater load carrying capability; the price paid is mechanical complexity and the need for providing means of work extraction and dissipation. The use of a piston engine for expansion into the two phase region is generally inefficient for most cryogenic fluids. Wet helium expanders are potentially attractive because the density difference between liquid and gaseous helium is relatively small and at liquid helium temperatures, the piston/cylinder materials have very little heat capacity; wall-gas heat transfer due to piston motion is minimized as a loss mechanism.

Thermodynamically, a two phase expansion engine always improves performance; a 0% isentropic efficiency corresponds to an isenthalpic expansion (i.e. a J-T valve). Johnson, Collins, and Smith (1), have reported test results on a two phase helium piston expansion engine that achieved approximately 90% isentropic efficiency. Figure 2 is a plot of the performance achieved by use of such an engine (i.e., 90% isentropic efficiency). It is similar to Figure 1 except that the ordinate is the ratio of refrigeration capacity of a system with a J-T valve to that with a wet engine. Alternately this can be viewed as flow required with a wet engine ratioed to that with a J-T valve for a given heat load. Important conclusions that can be drawn from these results are:

- o A wet engine helps a thermodynamically poorly designed system (i.e. low ε , high T, high ΔT) more than a well designed one.
- o Substitution of an engine with 90% isentropic efficiency for a J-T valve will result in at least a 67% increase in load carrying capability.
- o For a given load requirement, it is possible to operate a refrigeration system having an expander with smaller heat exchanger and less precooling, that is, lower ε and higher T.

SATURATED VAPOR COMPRESSION

By compressing the returning saturated vapor at 4.2K from the load pressure to some intermediate pressure, the entire system can operate at three pressure levels - load (lowest), precooler return (intermediate), and precooler delivery (highest). Since the regenerative heat exchanger size is in large part determined by the low pressure stream properties, an increase in this pressure can decrease size significantly. Minta and Smith (2) have presented cycle optimization data for one such system that can be operated at load temperatures as low as 1.8K without excessively large heat exchangers. Also, the input power required to compress the stream at ambient conditions may be decreased because of the cold precompression.

As a first step in evaluating cold compression, the J-T loop analysis was extended to include a 90% isentropic efficiency saturated vapor compressor. Because the compression process raises the temperature of the gas entering the low pressure side of the J-T heat exchanger, the high pressure gas also is warmer as it enters the final expansion process. Therefore, an expander is required to effect the final expansion since a warmer inlet to a J-T valve rapidly decreases liquid yield. Figure 3 is a performance map of the modified loop with a 90%

isentropic efficiency assumed for both expander and cold compressor and an assumed cold compression ratio of 2:1. The performance is plotted as in Figure 2, that is, the figure of merit is refrigeration capacity with J-T valve alone ratioed to that with both expander and compressor as a function of the variables ε, T and ΔT. Important results are:

- Relative to Figure 2, the curves are shifted down approximately 20% - this may be interpreted as a 20% decrease in input power for a given load. This may also be interpreted as a decrease in entropy generation (proportional to $\Delta P/P$) in the low pressure stream, with attendant efficiency increase.

- The performance curves are also shifted significantly to the left so that the same increase in load carrying capability can be achieved at much lower J-T heat exchanger effectiveness.

COLD EJECTOR

An ejector is a jet pump that utilizes excess flow in the main refrigerator to lower the pressure over the load by pumping on it (3) (4). The substitution of an ejector for the J-T valve has the advantage of increasing return stream pressure, as above with the cold vapor compressor, but with no pistons, valves or moving parts. One possible scheme is shown in the schematic of Figure 4; after cooling in the J-T heat exchanger, cold, high pressure gas enters the ejector at the primary inlet and is accelerated to approximately sonic conditions at the nozzle throat. This high velocity jet entrains a low pressure secondary flow that enters at the suction port, connected to the load.

The two streams exchange momentum in the mixing chamber and exit via a diffuser at some intermediate temperature and pressure. Part of the flow (secondary) proceeds to the load and part (primary) returns to the precooler via the regenerative heat exchangers. For given condition at the inlet to the primary nozzle the ejector improves on a J-T valve by using part of the pressure to do work on the secondary stream rather than to produce turbulence.

The key parameters in ejector performance are the mass entrainment ratio, $\mu = M_2/M_1$, and secondary flow pressure ratio, P_2/P_3. The conditions at the exit of the ejector are of particular importance to this ejector arrangement because they determine whether a heat exchanger is necessary in the secondary loop. If the secondary flow stream at a prescribed intermediate pressure does not have the capacity to produce refrigeration at 4.2K, it is necessary for the flow to pass through a recuperative heat exchanger before undergoing isenthalpic J-T expansion to the low load pressure.

Figure 5 is a performance map of a J-T loop with an ideal ejector, that is, one with no losses due to turbulence or friction (obviously a non-conservative approximation given the mixing process). The ratio of the power required to compress the low temperature loop flow with an ejector and with a J-T valve (at the same inlet conditions) is plotted as a function of secondary compression ratio for various recuperative heat exchanger pinches and fixed temperature level of 17K. Each point on the curve corresponds to a particular mass entrainment ratio μ (<1.0).

Important qualitative results from this plot are:

- For the ejector to be advantageous, small pinches must be maintained on the final regenerative exchanger.

- For high secondary flow compression ratios and small pinches, a heat exchanger will be necessary in the secondary loop.

The actual combination of entrainment ratio and secondary compression ratio results from the ejector performance. Differences between this performance and that assumed will further limit the applicability of Figure 5.

SUMMARY AND CONCLUSIONS

The final expansion stage of typical liquid helium temperature refrigerators consists of heat exchange followed by isenthalpic expansion through a J-T valve. This highly irreversible process is a large source of entropy generation which may be reduced in at least three ways to lower the input power required per unit of refrigeration:

- Final expansion through a nearly isentropic work extraction device will increase refrigeration capacity upwards of 67%.

o Combination of this expander with a saturated vapor compressor will further increase capacity approximately 20% and decrease total system entropy generation by increasing return flow pressure level.

o Substitution of an ejector or jet pump for a J-T valve has the potential for increasing refrigeration capacity and decreasing losses (similar to cold vapor compression) with a device having no moving parts.

LITERATURE CITED

1. Johnson, R.W., Collins, S.C., Smith, J.L. Jr.: "Hydraulically Operated Two-Phase Helium Expansion Engine"; Advances in Cryogenic Engineering, Vol. 16, pp 171-177 (1970).

2. Minta, M., Smith, J.L. Jr.; "Helium Liquefier Cycles with Saturated Vapor Compression"; to be published in Advances in Cryogenic Engineering, Vol. 27.

3. Daney, D., McConnell, P.M., Strobridge, T.R.; "Low Temperature Nitrogen Ejector Performance", Advances in Cryogenic Engineering, Vol., 18, pp 476-485 (1972).

4. Rietdijk, J.A., "The Expansion Ejector, A New Device for Liquefaction and Refrigeration at 4K and Lower", Annex 1966-5 Bull. I.I.R., pp. 241-249, Com. I, Boulder.

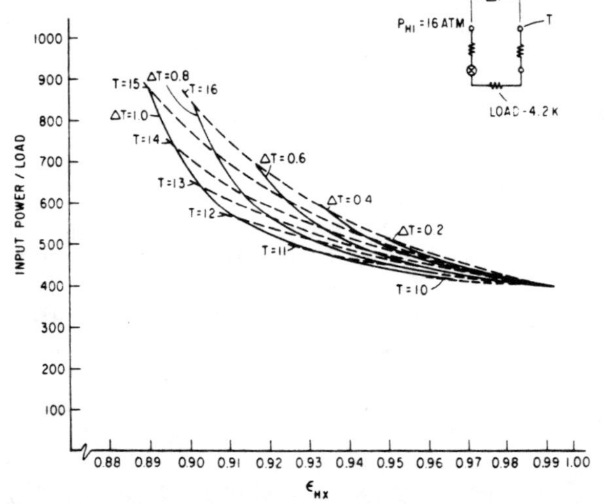

Figure 1. Thermodynamic performance map-j-t loop with valve.

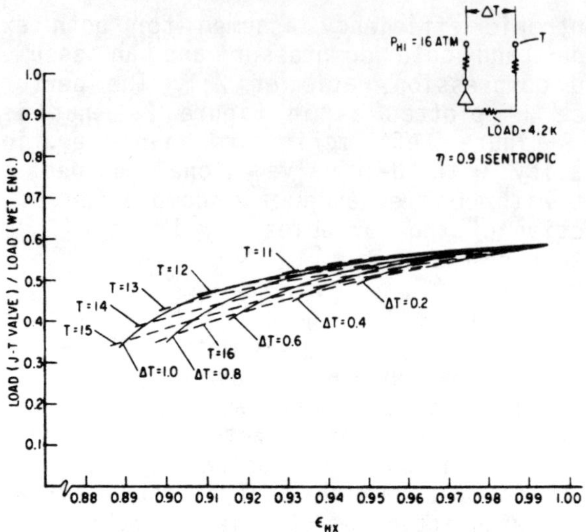

Figure 2. Thermodynamic performance map-j-t loop with wet engine.

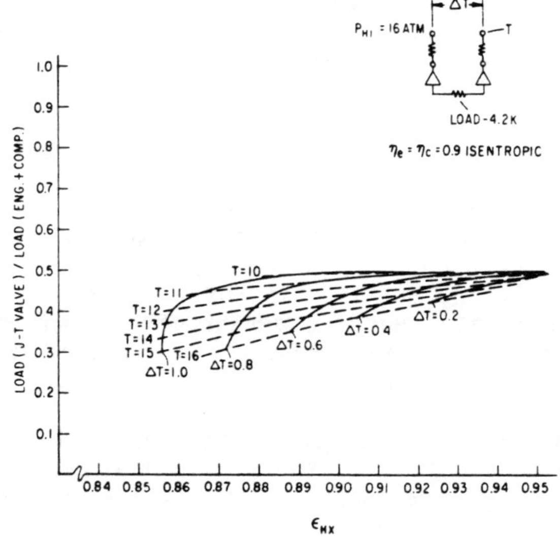

Figure 3. Thermodynamic performance map-wet engine & cold compressor.

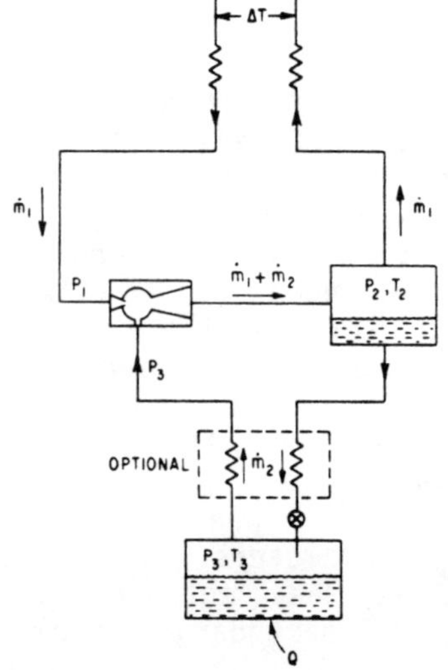

Figure 4. Ejector/j-t loop schematic.

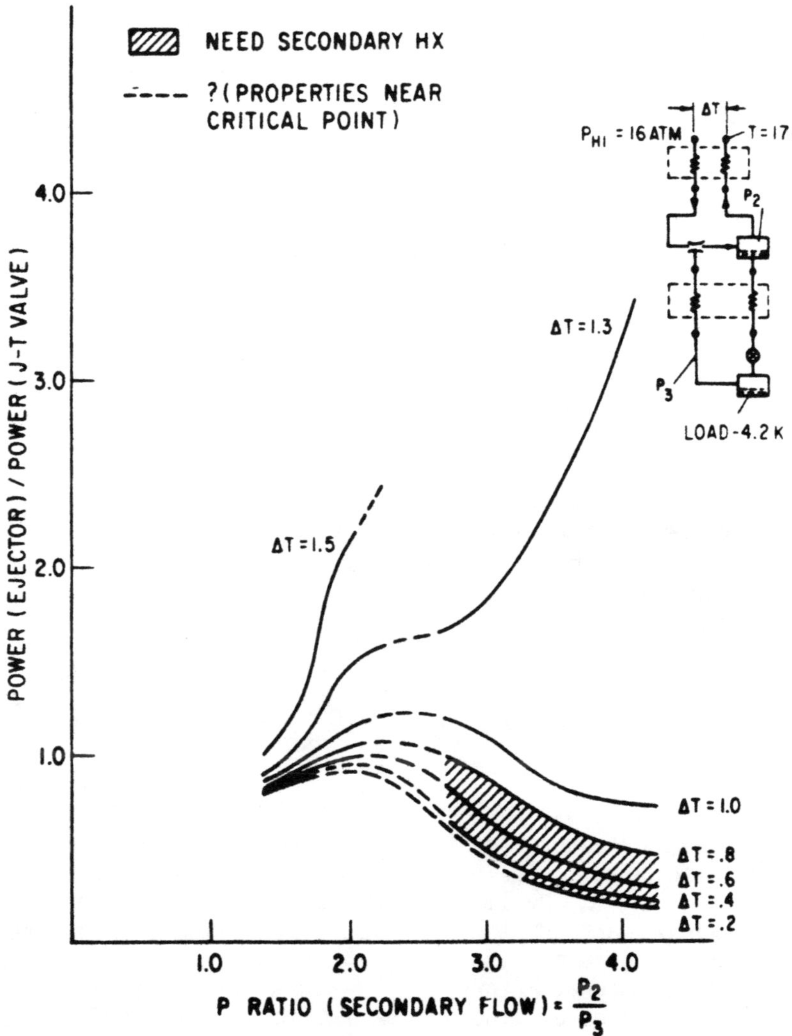

Figure 5. Ejector thermodynamic analysis.

LARGE SCALE SUPERFLUID PRACTICE

S. CASPI
C. TAYLOR
W. S. GILBERT
W. HASSENZAHL
J. RECHEN
and
R. WARREN

Accelerator and Fusion Research Division
Lawrence Berkeley Laboratory
University of California
Berkeley, California 94720

In the past three years we have been testing superconducting dipole magnets in superfluid He II. Superfluid is produced on a continuous basis at 1 atm pressure and in quantities of up to 400 liters. The cryostat can be used as a large calorimeter because of the unique properties of He II. This paper reviews refrigeration design and operation and describes the way in which our magnets behave in superfluid.

INTRODUCTION

Since 1979 Lawrence Berkeley Laboratory has been testing superconducting magnets in He II. The 1 atm pressure, 1.8 K, He II, test facility, is an integral part of the LBL Research and Development program on high field superconducting dipole magnets for particle accelerators [1]. Some of the experience gained in this facility and the details of its operation are reported here.

EXPERIMENTAL TEST FACILITY

The dewar is based on the principle of the Claudet bath [2] and provides He II at 1.8 K and 1 atm on a continuous basis. We report here on the dewar in the vertical configuration [3], Fig. 1. The facility was modified recently to accommodate horizontal magnets and a He II volume of up to 400 liters [4].

The vertical dewar consisted of a 28-ℓ He I chamber, a 142-ℓ He II chamber, and a He II refrigeration system. The He I chamber is a heat intercept for the magnet current leads and instrumentation wires, a liquid supply for 1.8 K refrigeration, and an atmospheric pressure intercept for the lower He II reservoir. The tube connecting the two chambers permits mass flow to maintain atmospheric pressure in the lower vessel. During steady state operation, He I at 4.4 K and ~1 atm is precooled to about 2.6 K in a counter flow heat exchanger. It then expands through a regulated Joule Thomson valve, and exits at a vapor pressure corresponding to about 1.75 K. Downstream, it exchanges heat with the 1.8 K He II reservoir, then precools the counterflow heat exchanger, and finally exits to a pump.

Seven carbon glass thermometers were placed inside the dewar. In the upper vessel, T2 and T1 were mounted 1 and 10 mm from the bottom of the upper vessel respectively. In the lower, He II, reservoir, the sensors were mounted as follows (distances are below the top flange of the lower vessel): T3 - 5 mm, T4 - 30 mm, T5 - 365 mm, T6 - 835 mm, and T7 - 1225 mm at the bottom of the dewar. The temperature was measured with an accuracy better than 5 mK below T_λ using an H.P. 9845 data acquisition system.

COOLDOWN AND PRODUCTION OF He II

During a typical cooldown the liquid in the region of the heat exchanger reached T_λ (see Fig. 2) approximately three hours

*This work was supported by the Director, Office of Energy Research, Office of High Energy and Nuclear Physics, High Energy Physics Division, U. S. Dept. of Energy, under Contract No. DE-AC03-76SF00098.

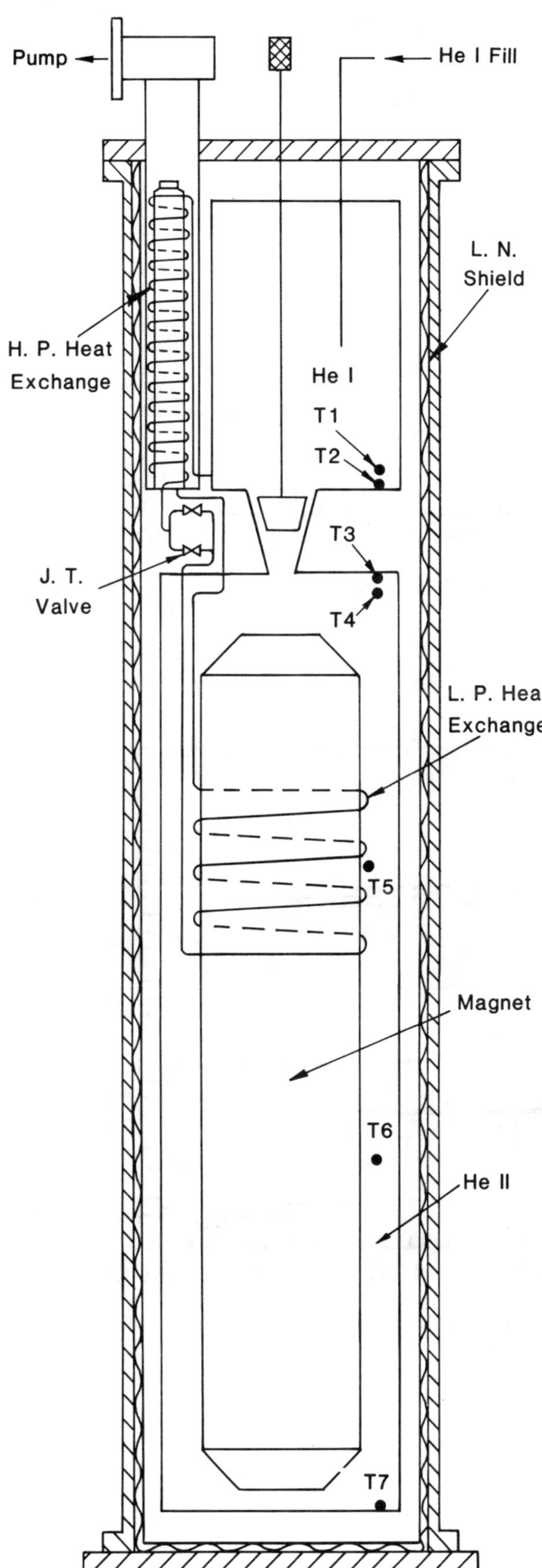

Figure 1. Helium II refrigeration system for testing superconducting magnets.

after the JT valve was set into operation. During this period, the liquid below the heat exchanger showed good mixing resulting in a temperature difference T7 - T5 < 100 mK. Above the heat exchanger, the liquid was stagnant and the temperature remained near 4.2 K. Temperature sensor T5 was the first to reach T_λ and remained at T_λ until the rest of the liquid in the lower reservoir was cooled to T_λ. The expansion of this "lambda liquid" is gravitationally free and is caused by the heat transfer process at the He I He II boundary. Helium II expands upwards and downwards but at different rates because of the different bath temperatures in the two directions. Experimentally when the transition boundary passes a sensor location the measured temperature drops to T_λ and remains constant. The boundary is sharp, sensor T3 does not see the cold boundary approaching until the temperature at T4, which is only 25 mm away, has been at T_λ for some time. We estimate T3 remains around 4 K until the boundary is within 1 mm of the sensor. The boundary is thought to be even sharper than the 3 mm thickness of the sensors. The boundary velocity between sensors T6 and T7 which start at $T_\infty = 2.2$ K is 2 mm/sec and between sensors T3 and T4 which start at $T_\infty = 4$ K, is 0.1 mm/sec. These values are typical but depend on the refrigeration power. An estimate on the thickness of the He II - He I boundary layer is reported in Ref. (5).

JOULE THOMSON (JT) EXPANSION VALVE

The temperature across the JT valve during cooldown is shown in Fig. 3. The upstream temperature depends on the heat exchange between the incoming He I liquid and the return cold vapor through the counterflow heat exchanger. The downstream temperature usually reflects the equilibrium between temperature and pressure according to the saturation curve. During the initial cooldown period however, for a pressure less than 40 Torr, the downstream temperature remains at ~2.2 K regardless of the pressure. This behavior changes when the lower reservoir drops below 2.6 K. At this time the downstream temperature falls abruptly to its equilibrium value of 1.8 K and any change in pressure is immediately reflected in a change in the saturation temperature. Simultaneously the upstream temperature levels off about 2.6 K and, as long as this value is maintained, the overall operation is stable. When the

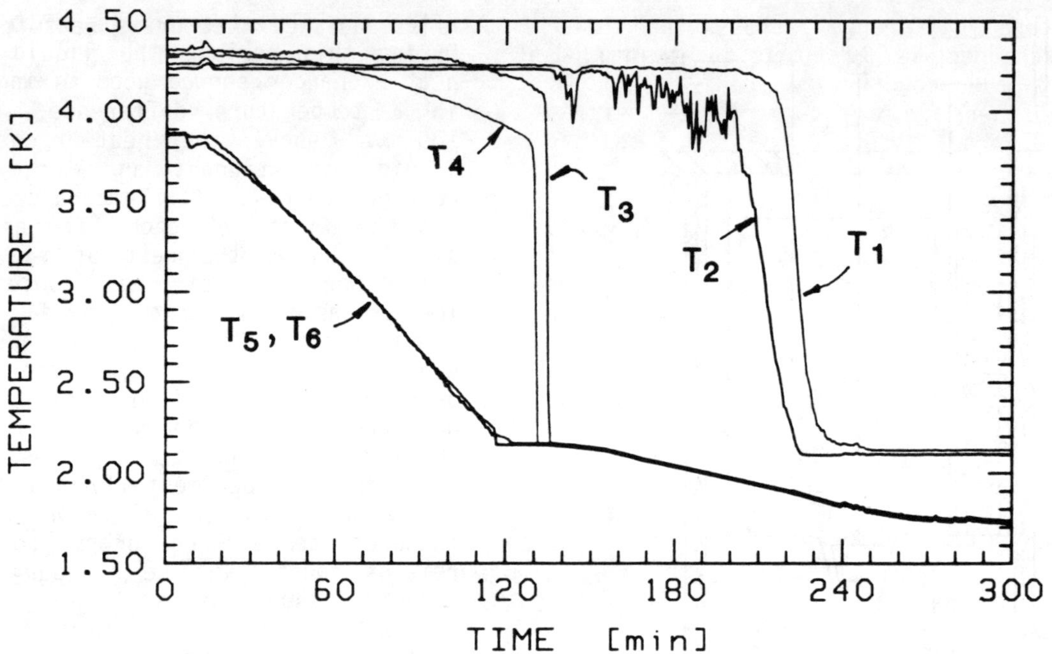

Figure 2. Temperature at various locations in the cryostat during cooldown.

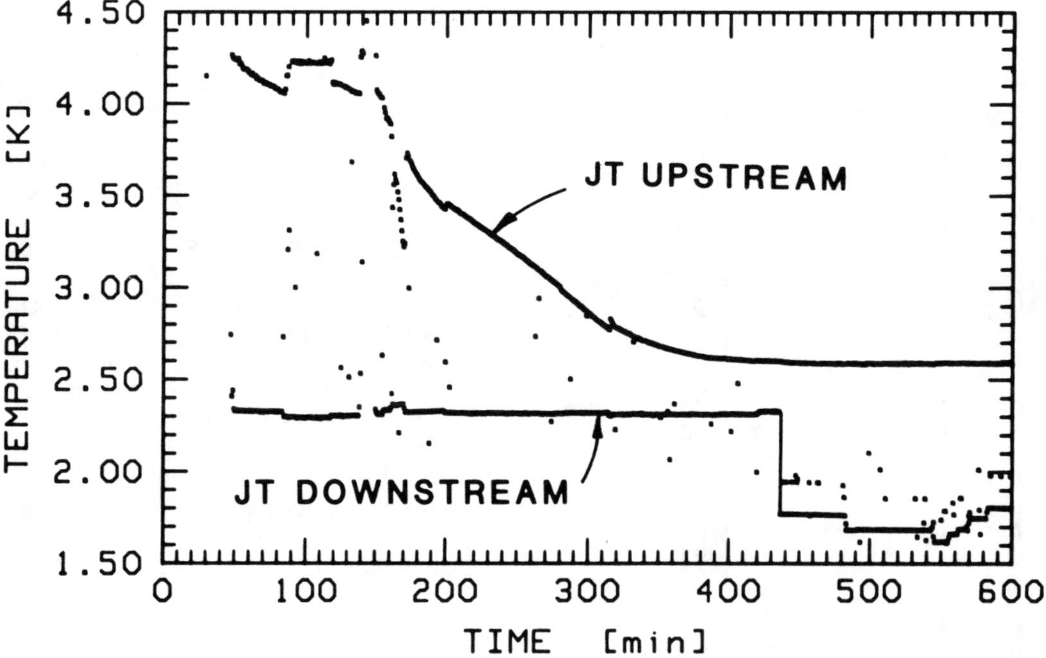

Figure 3. Temperatures at the JT valve during cooldown.

temperature difference across the lower heat exchanger is small (e.g. less than ~100 mK) the counterflow heat exchanger cools further and the temperature upstream of the JT drops abruptly to ~2 K. At this temperature the liquid behaves as superfluid and it flows unimpeded through the JT valve. The uncontrolled rush of liquid floods the lower heat exchanger, reduces the cooling efficiency and as a result the system develops a thermal instability. When this happens the JT valve must be shut off so that the heat exchanger dries out and the upstream liquid warms up to He I temperature before cooling can resume.

An undesirable thermal condition can develop under similar circumstances where there is excess refrigeration. When the temperature of the lower reservoir is too low (<1.7 K) or the temperature difference across the heat exchanger is below ~100 mK, the heat flux through the channel connecting the lower and upper vessels is reduced. Accordingly the temperature gradient across the channel is reduced by increasing the effective length from the He II/He I interface to the main He II bath. As result the interface moves up through the tube until it crosses the channel entrance at the bottom of the He I reservoir. A stable layer of cold helium with a temperature $T_{\lambda}-\varepsilon$ is established that draws heat from He I by conduction only (Fig. 2).

SUPERLEAK

One of the design goals of the He II test facility was to have a simple procedure for magnet installation before and after each test. This was accomplished by using a breakable seal between all flanges. The application of an epoxy resin seal (50-50 mix of Shell Epon 828 and Versamid 140) was found to be leaktight even though superleaks can develop in He II. This procedure has been used about 30 times and found to be reliable. The only time a superleak was observed was when the membrane of a pressure transducer developed a superleak. The recorded superleak is shown in Fig. 4 and described below.

The cooldown from 4.4 K proceeded at first with no indication that any leak was present, as indicated by cooldown rate and vacuum. After superfluid was created and propagated throughout the dewar the He II/He I interface finally reached the top of the He II vessel and entered the connecting tube. At this time the superfluid also reached the superleak in the pressure transducer located on the upper flange. The superleak spoiled the vacuum and the gas in

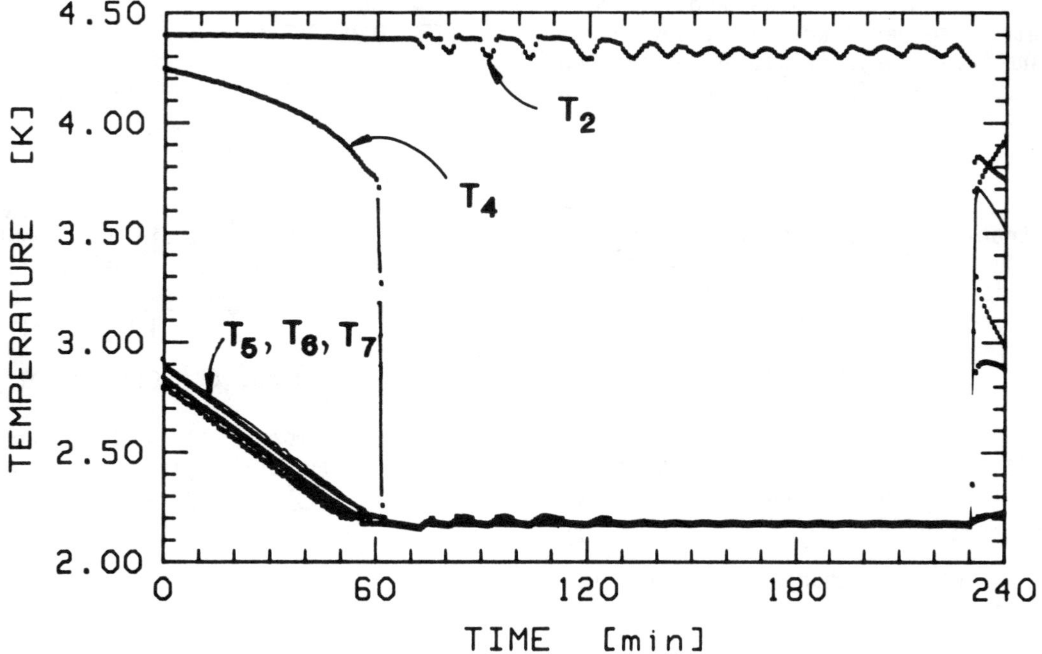

Figure 4. A superleak inhibits cryostat cooldown below T_{λ}.

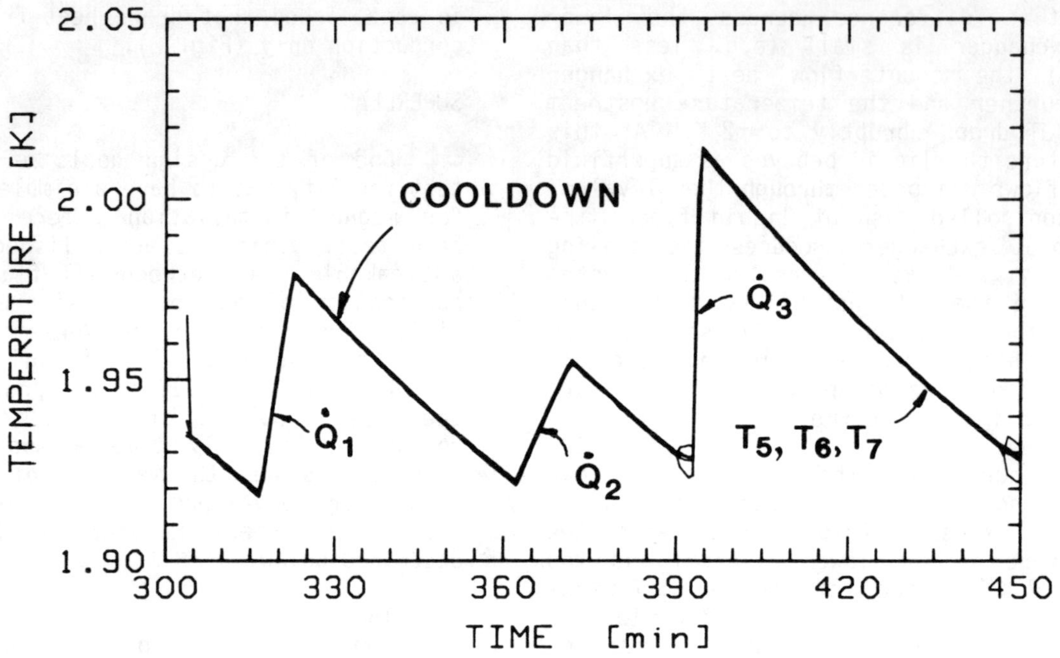

Figure 5. Superfluid temperature response to various heat inputs.

the vacuum space cooled the He I vessel and warmed up the lower dewar by convection. This heat leak raised the temperature in the vicinity of the superleak above the lambda temperature and the flow of helium into the vacuum stopped. Slowly the vacuum was pumped to its original value and cooldown resumed at a temperature just above T_λ and continued until superfluid again reached the superleak area. This cycle was recorded for over 3 hours with the lower reservoir temperature oscillating around T_λ. In the time sequence the behavior indicated where the superleak might be and when the pressure transducer was removed the superleak went away. To find such a superleak at room temperature is quite difficult if not impossible and this string of events led to its elimination.

CALORIMETRY

The isothermal behavior of superfluid He II and the absence of vapor when it is used at 1 atmosphere provide the means for calorimetry using a straightforward energy balance. The rate of change of temperature during magnet cycling is plotted in Fig. 5 and temperature jumps due to energy dumps during magnet quenches are shown in Fig. 6. The absence of stratification is clearly visible although some of the temperature sensors are located as far apart as 1.5 m.

LITERATURE CITED

1. Taylor, Clyde., Althaus, Robert., Caspi, Shlomo, Gilbert, W.D., Hassenzahl, W.V., Meuser, R.W., Rechen, Jeb, Warren, R.P. 'Design of epoxy-free superconducting magnets and performance in both Helium I and pressurized Helium II.' IEEE Trans. on Magnetics, Vol. MAG-17, No. 5, 1981, pp. 1571-1574.

2. Bon Mardion, G., Claudet, G., Seyfert, P., Verdier, J. 'Helium II in low temperature and superconductive magnet engineering.' Adv. Cryo. Engineering, Vol.23, pp.358-362.

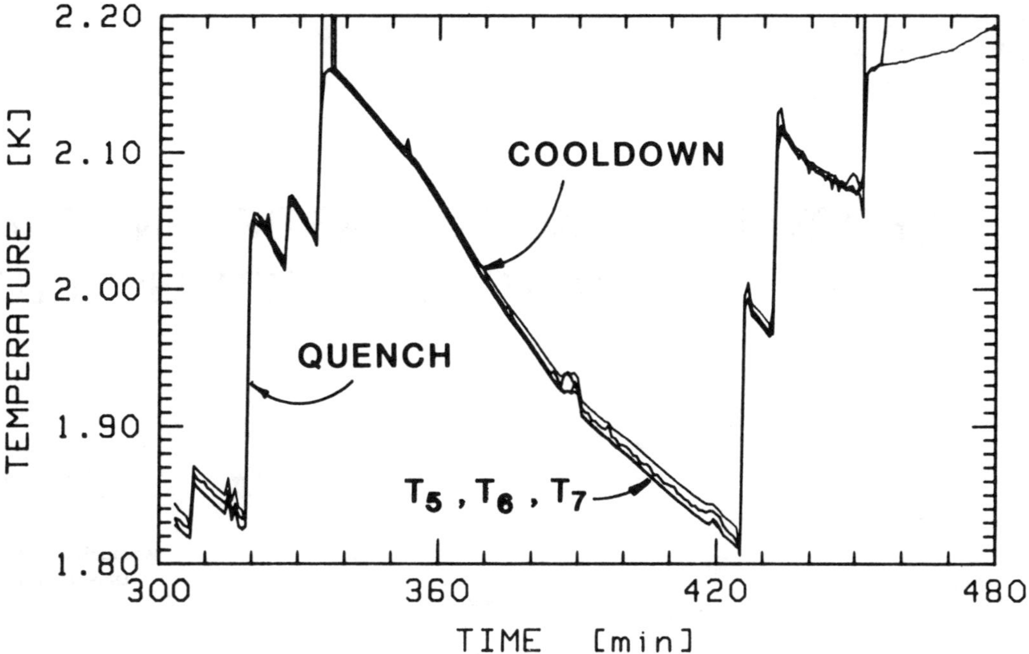

Figure 6. Cryostat temperature variations for several quenches with some cooling in between.

3. Warren, R. P., Lambertson, G. R., Gilbert, W. D., Meuser, R. W., Caspi, Shlomo, and Schafer, R. V. "A pressurized Helium II - cooled magnet test facility." 8th Inter. Cryo. Engr. Conf., (1980), pp. 373-377.

4. Caspi, Shlomo. 'The use of calorimetry in superfluid He II to measure losses in superconducting magnets.' 9th Inter. Cryo. Engr. Conf., Kobe (1982).

5. Caspi, Shlomo. 'Gravitational convection of subcooled He I and the transition into superfluid He II at atomospheric pressure.' 8th Inter. Cryo. Engr. Conf., (1980), pp. 238-242.

A GROOVED SELF-ACTING GAS BEARING FOR USE IN CRYOGENIC EXPANSION TURBINES

K. J. YANG

Cryogenics Laboratory
Chinese Academy of Sciences, China

A. J. MUNDAY

Mechanical Engineering Department
University of Southampton, U.K.

The application of cryogenics and superconductivity requires reliable and compact helium liquefiers or refrigerators, and the miniature high speed expansion turbine is the heart of this kind of liquefier-refrigerator. It must be miniature because even a moderate 'cold' production requirement with cold high-density gas results in small gas through-ways, and it must rotate at very high speed in order to get optimum performance. Compared to reciprocating expansion engines, the advantages of expansion turbines are: small size for large flow capacity, low vibration and noise, and using gas bearings, without contamination and low wear so that they can work for a very long time.

For many years, considerable work has been done to develop a suitable gas bearing for cryogenic expansion turbines but the results are not completely satisfactory. The requirements for gas bearings in cryogenic expansion turbines are:

1) reliable operation at very high speed (for instance 200,000 ∿ 600,000 rpm) without half speed whirl instability;

2) small wobble of the rotor centre in order to design a small clearance between the turbine wheel and the nozzle ring so that a higher efficiency can be assured;

3) small amount of heat generated in the clearance space of the bearing and a method by which the heat can be easily transferred;

4) accomodation to temperature variation.

It is anticipated that the new cryogenic machine with gas bearings will save power and reduce manufacturing costs.

There are several types of high speed gas bearing which can be considered:

a) externally pressurized,

b) self-acting
   i) lobed,
   ii) axially grooved,
   iii) spiral gorvved,
   iv) foil,
   v) tilting pad.

The problems with externally pressurized bearings are the requirement for external compressor power, and their complex geometry. Of the self-acting bearings, the existing foil bearings do not allow only a small wobble of the rotor, and the tilting pad bearings also have a complex geometry. The most stable of the other bearings is the spiral groove bearing and in the form of a herringbone grooved bearing it can accommodate the largest load capacity.

The herringbone journal bearing is the most suitable choice for cryogenic applications, and the experience with this type of bearing for a Freon expansion turbine at the University of Southampton makes the development of this new machine easier.

Vohr and Chow (1) theoretically investigated a herringbone grooved journal bearing in 1965. They found that the load capacity of a herringbone grooved bearing increases without limit with increase in rotation speed, and it can inhibit the notorious half-speed whirl instability. Fleming and Hamrock analyzed and determined the optimum groove parameters to maximize the radial load capacity in 1974 (2). After that, in 1976, they went a step further and determined the optimum groove parameters to maximize the bearing's stability (3) by using the analysis of Vohr and Chow. So, a very good foundation has been laid for design and experiment.

This paper describes the design of and experiments with a test machine with herringbone grooved bearings for the working conditions of cryogenic turbines. In these experiments the stability of herringbone grooved gas bearings at very high speed has been observed and thus the possibility of their use in cryogenic expansion turbines has been demonstrated.

DESIGN AND CONSTRUCTION

Simulation of a Cryogenic Expansion Turbine

In order to reduce the cost of machining and for convenience of doing different experiments, an experimental machine, instead of a real expansion turbine, has been designed and fabricated. Its structure is shown in Figure 1. The diameter of the shaft was chosen at 7mm. Three capacitance probes, A, B and C, are fixed at the top, the middle and the bottom of the shaft, separately measuring the shaft vibration. For convenience in measuring the wobble of the turbine wheel which affects the efficiency of the turbine, a 'false' turbine wheel, a small aluminium cap, is fixed on the bottom of the shaft. The nozzle ring is mounted at the top of the shaft, instead of the bottom, so that the pressurized gas drives the blower wheel, which would normally be used for braking and exporting power in the cryogenic expansion turbine.

A simple externally pressurized gas thrust bearing is mounted at the bottom of the rig for supporting the shaft weight. The area of the cross-section of the 'shaft neck' is reduced as far as possible to prevent the 'cold' from transferring upwards. Because of the same reason, the wall of the rig housing is designed to be very thin at this position. The cooling water has two functions. It can carry the heat away when the heat is generated in the bearing clearance and it also can keep the bearing warm if too much 'cold' is transferred from the bottom.

The shaft strength has been calculated very carefully. Three factors, high speed rotation, interference fit and different material contraction at low temperature, have been considered in the calculation.

Herringbone Grooved Bearings

The construction of a herringbone grooved bearing is shown in Figure 2. There are some angled shallow grooves in the journal surface. The grooves can be partial, as shown, or extend the complete length of the bearing. Also, the grooves can be placed in the rotating or non-rotating surface. When one of the bearing surfaces rotates, the groove acts as a viscous pump pumping fluid toward the centre of the bearing thereby increasing the lubricant pressure.

Fleming and Hamrock determined groove parameters to maximize the bearing's stability by using the Newton-Raphson method and the method of vector analysis. Because the bearing's stability is very sensitive to changes in groove parameters and manufacturing tolerances must be allowed, the dimensionless stability parameter M of the bearing to be designed should be less than the minimum theoretical stability limit M_C. For the bearings with grooved member rotating at high bearing compressibility number Λ, the safety coefficient is 2 or more.

The present design (4) selects the bearing with grooved member rotating. The dimensions of the shaft are shown in Figure 3 and the material of the shaft is stainless steel (EN58AM). In the design the rotation speed of the shaft required is 400,000 rpm.

Generally speaking, by selecting smaller radial clearances, higher load capacities and stability can be obtained. But, with reduction of the clearance, machining becomes more difficult and the viscous heat due to rotation generated in the bearing clearance will increase. Thus the clearance was chosen to give only the required operating speed to minimize bearing friction losses. Because the gas pressure at the exit of nozzles in cryogenic turbines is usually higher than 1 atmosphere, calculations for both 1 atm and 3 atm have been done. The calculated results for the two different pressures are only very slightly different. According to the analysis, a herringbone grooved gas bearing designed for air would be more stable when helium is used for lubrication. Thus the present design is produced using air parameters so that initial test work may be done using air while the final arrangement using helium will have an added safety factor.

By using the optimization of Fleming and Hamrock and considering the shaft growth at high rotational, speed, the following bearing parameters have been decided upon for machining:

Radial clearance	h_r	= 7 μm
Number of grooves	N_0	= 10
Groove angle	β	= 34.1°
Width of groove	b_g	= 1.32 mm
Depth of groove	d_g	= 10 μm
Length of groove	L_1	= L = 7mm

Construction

One may think that the construction of a high speed herringbone grooved gas bearing could be very difficult. But in fact, for a miniature expansion turbine with herringbone grooved bearings, because of their very high stability, the bearing clearance required is larger than that for other self-acting bearings.

When the bearing bore was produced, a general lathe was used first and then it was lapped by using very fine lapping paste. The outer circular surface of the shaft was produced by an accurate cylindrical grinder, and then the grooves were etched by nitric acid. Finally, the shaft was balanced as accurately as possible and the residual unbalance at the bottom of the shaft is approximately 0.54 g.mm/kg. Figure 4 shows the photograph of the shaft with herringbone grooves and beside the shaft there is a mask used when etching the grooves into the shaft.

EXPERIMENTAL PROCEDURES AND RESULTS

Bearing Stability and Vibration Level

In order to measure the bearing stability and the vibration level, three small capacitance probes with three Wayne Kerr vibration meters were used. The probes were calibrated carefully. An infrared ray pickup and a digital frequency meter were used to measure the rotation speed, and a four trace oscilloscope was used to observe the vibration and the speed of the shaft.

According to the calculation of the shaft strength, when the shaft rotation speed is 400,000 rpm, the maximum stress at the centre of the top turbine wheel has reached the yield stress of Austenitic Stainless Steel. Therefore during the experiments the shaft was usually allowed to rotate only up to 400,000 rpm, but the shaft once rotated up to 420,000 rpm when helium was used for lubrication and driving, and no problems occurred.

The experimental results show that the rotation of the shaft is very stable and no half-speed whirl has been detected. The amplitude of the vibration of the shaft is very small. At the top, it is only 2 μm pk - pk; at the middle of the shaft, it is even less than 1 μm pk - pk, and it almost does not increase with increasing shaft speed. The amplitude of the vibrations at the bottom of the shaft is slightly bigger, and it slowly increases with the shaft speed. This is possibly because the neck of the shaft is too thin so that the centrifugal force makes the neck bend slightly.

For many times the shaft has been suddenly started up to 400,000 rpm, or stopped by quickly turning off the driving gas. Most of the experiments have been with vertical axis operation which would be the normal position for the expansion turbine. The machine has, however, been tested with an inclined axis to see the effect of a radial load on the journal bearings. An impact test has also been performed. The rotation has always been very stable, and the amplitude of the vibration has not permanently changed.

Viscous Friction Power

Because of the viscous friction between the shaft and the gas in the bearing clearance, there is some heat generated. Approximate measurements of the bearing temperature rise give only 23°C at 400,000 rpm even when no cooling water is used, so this would have no serious consequences for the materials used or heat flow to the turbine.

It was quite difficult to measure the bearing friction power exactly. By using a dynamometer to measure the friction force, and by measuring the deceleration of the shaft when the driving gas was quickly turned off, the bearing friction power was measured to be approximately 30 watts for the bearing viscous friction at 400,000 rpm. But the calculated value is 17 watts. So, there are some doubts regarding accuracy of the experimental measurements.

Low Temperature Experiment

Figure 5 shows the low temperature experiment. The bottom of the rig has been cooled by liquid nitrogen, and the helium feeding the bottom thrust bearing has been pre-cooled in liquid nitrogen. When the shaft rotated at 400,000 rpm, the temperature of the rig bottom was -167°C and the bearing temperature was 27°C; no cooling water was used and the ambient temperature was 18°C.

CONCLUSIONS

It can be seen from the experimental results that the advantages of using herringbone grooved gas bearings in cryogenic expansion turbines are:

a) can stably rotate at high speed;

b) relatively simple to manufacture;

c) very small amount of shaft wobble; and small amount of heat generated;

d) uses the same pressure in the bearing as at the exits of the nozzles, so no labyrinth seal is needed.

Together with the fundamental advantage of a self-acting bearing of not requiring the power of an external compressor, it can be concluded that the herringbone spiral groove bearing has many advantages which make it suitable for use in cryogenic expansion turbines.

FURTHER EXPERIMENTS

Further experiments will include measuring the bearing vibration capacity and ultimate stable speed. Also, it is necessary to know what is the lowest temperature which the bearing can operate. A spiral groove thrust bearing could be produced in order to remove completely the external gas supply from bearings.

ACKNOWLEDGEMENTS

We would like to thank Mr. R. W. Woolley for all his help in the design and experiments, and Mr. P. Humby for his skilled and patient work in building and making modifications to the machine.

LITERATURE CITED

1. Vohr, J.H. and Chow, C.Y. "Characteristics of Herringbone-Grooved, Gas Lubricated Journal Bearings" Trans. A.S.M.E., Sept. 1965.

2. Hamrock, B.J. and Fleming, D.P. "Optimization of Self-Acting Herringbone Grooved Journal Bearings for Maximum Radial Load Capacity", 5th Gas Bearing Symposium, Southampton University, March 1971.

3. Fleming, D.P. and Hamrock, B.J. "Optimization of Self-Acting Herringbone Journal Bearings for Maximum Stability", 6th Gas Bearing Symposium, Southampton University, March 1974.

4. Yang, K.J., "Design of an Experimental Device with Herringbone Grooved Gas Journal Bearings for use in High Speed Cryogenic Expansion Turbines", Report No: ME/82/8, Southampton University, May 1982.

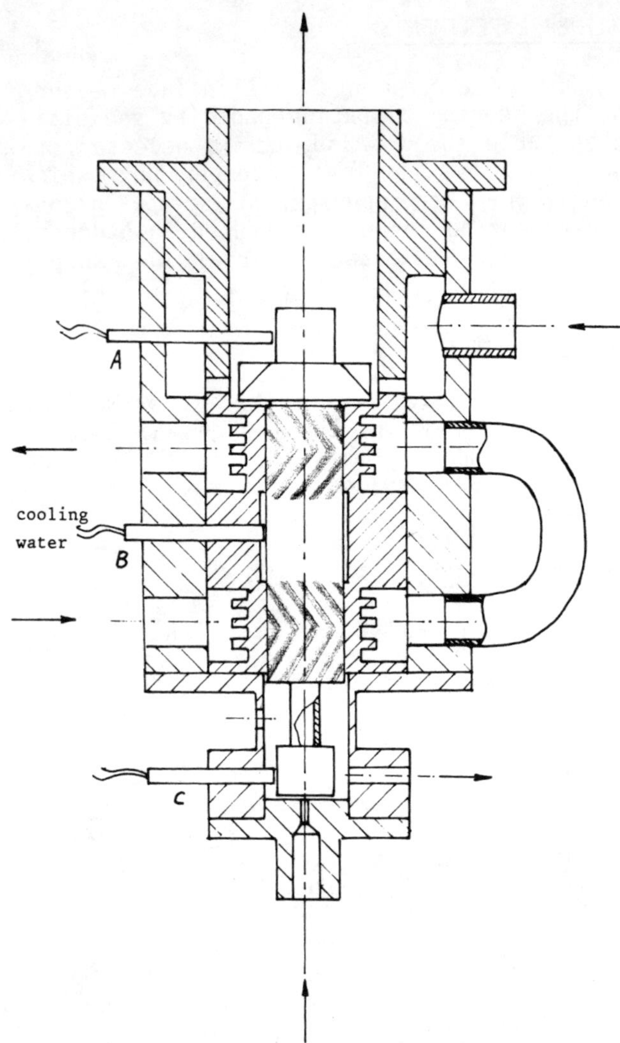

Figure 1. Experimental machine for the simulation of a cryogenic expansion turbine.

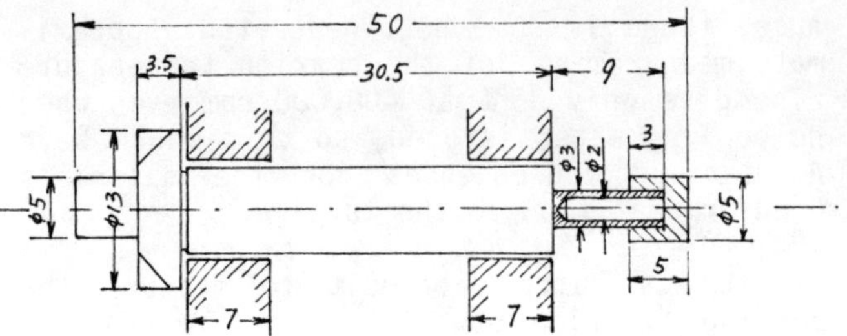

Figure 3. Shaft dimensions.

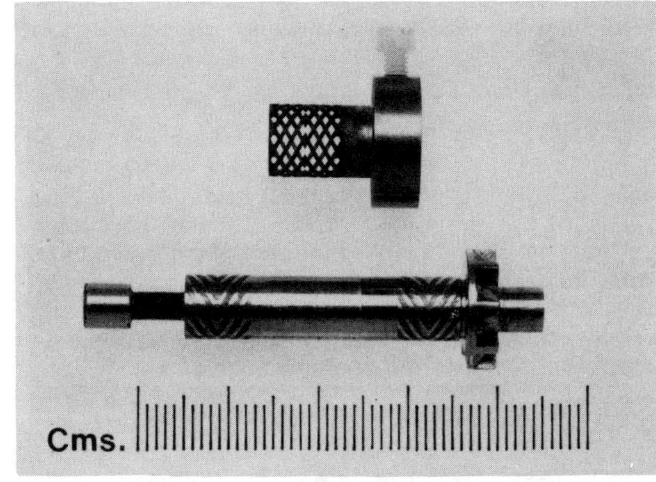

Figure 4. Photographs of the shaft with herringbone grooves and mask used for etching the grooves into the shaft.

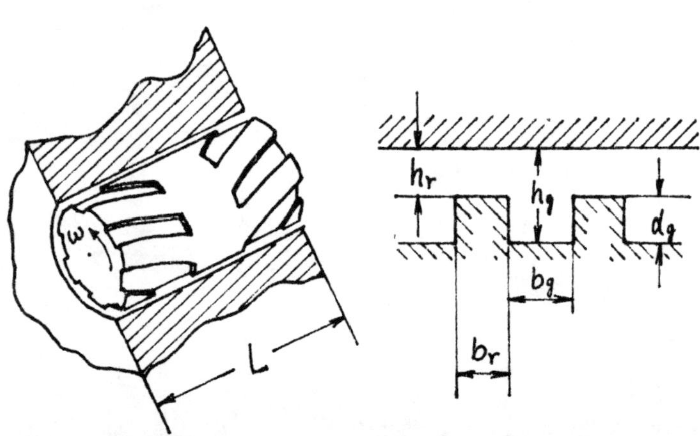

Figure 2. Construction of a herringbone grooved bearing.

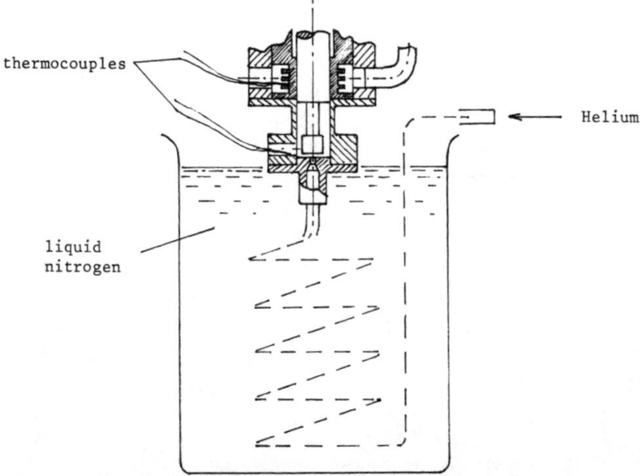

Figure 5. Schematic of the low temperature experiment.

COMPUTER CONTROLLED HELIUM EXPANSION ENGINE

The Navy is developing a 10 liter/hour helium liquefier as a part of its Superconducting Propulsion Machinery Development Program. The liquefier development effort is geared to fully automatic operation under steady state conditions as well as cooldown. A hydraulically operated expansion engine has been designed and fabricated at the David Taylor Naval Ship Research and Development Center so that one can control the variabilities of the expansion cycle through several electrical signals. A small computer has been programmed to control this expansion engine in various operating modes. Results of these experiments are presented.

G. PATTON
G. GREEN
K. DUNN
and
V. DILLING

Helium liquefiers which use reciprocating piston expanders have historically been designed with a crankshaft, flywheel, and connecting rod to achieve the reciprocating motion of the piston. This mechanism is very satisfactory for liquefiers in stationary or laboratory service, but the stresses which arise in this mechanism when the liquefier is exposed to a mobile environment make a reciprocating system with less mass and inertia more attractive.

The helium liquefier, whether used as a refrigerator or liquefier, is usually an ancillary device to cool a magnet or electronic sensor. The personnel associated with the device are probably not experts in cryogenics or helium liquefiers. It would be desireable, therefore, if a liquefier could be started, cooled down and operated automatically. The ability to adjust the refrigeration capacity and the operating parameters of the device by tailoring the engine performance to the thermal load will lead to a system which is more efficient.

In an attempt to achieve the desired automatic operation, a small Digital PDP 11 computer was programmed to control the variable functions of an electronically controlled, hydraulically operated, helium expansion engine. The logic scheme of the feedback network was defined and programmed for the computer. A series of preliminary experiments were conducted using a single expansion engine as a test rig and the results are presented.

EXPANSION ENGINE OPERATION

The reciprocating helium expansion engine follows the Brayton cycle, where the cylinder is partially filled with gas at constant pressure, expanded isentropically, (or nearly so) and exhausted at constant pressure. Figure 1 shows the pressure-volume (P-V) diagram for an ideal expansion process. Figure 2 shows a photograph of the actual pressure volume diagram generated by the helium expansion engine. The helium inlet and exhaust valves, shown schematically in Figure 3, must be timed and actuated to achieve this Brayton Cycle. Additionally, the "active accumulator"(1) which returns the piston must be pressurized and vented in time with the exhaust and expansion stroke.

Expansion engines presently use mechanical, mechanical/pneumatic, or mechanical/electric techniques to operate the helium inlet and exhaust valves. While these work well, they do not allow expansion cycle to be altered while the expansion engine is in operation.

In order to achieve the desired timing electronically, the precise location of the expansion piston is required by the electronic valve control logic. This position signal is derived from a linear variable differential

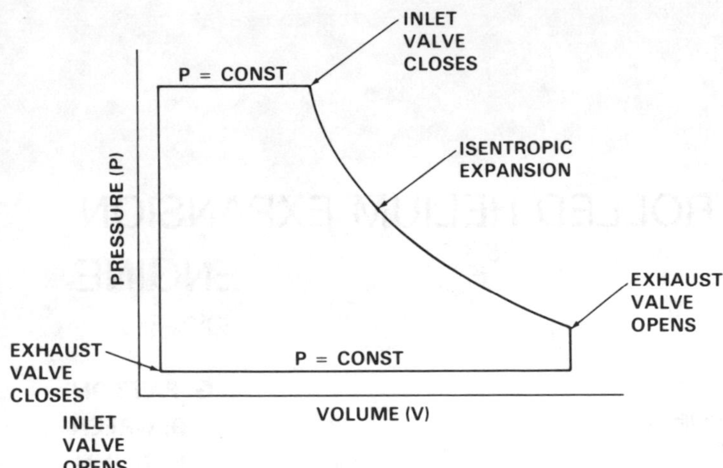

Figure 1. Pressure-volume (P-V) diagram for ideal expansion cycle.

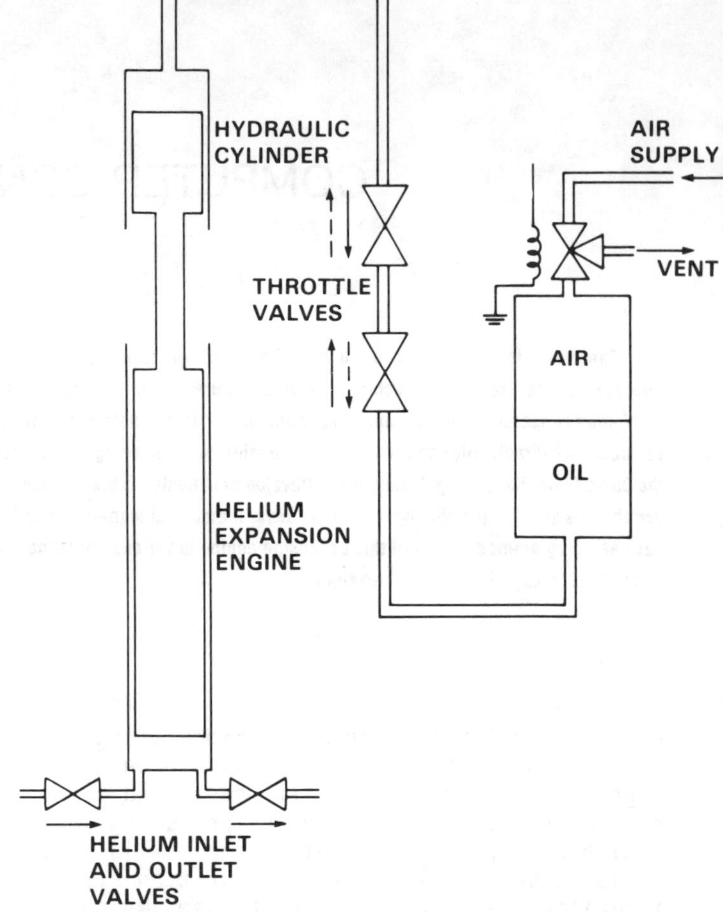

Figure 3. Schematic drawings of hydraulic expansion engine with active accumulator.

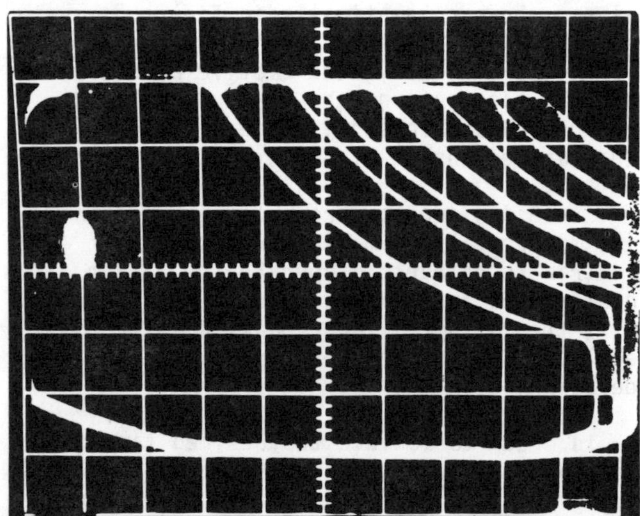

Figure 2. Photograph of actual pressure-volume (P-V) diagram.

transformer (LVDT) which is installed so that its core travels with the reciprocating piston assembly. The transformer position signal is compared to a preset timing voltage in an electronic comparator. Figure 4 shows a block diagram of the helium valve controls with potentiometers supplying the control voltage. This comparator output sets or resets a flip-flop to control a solenoid air valve that supplies air to the helium valve operator to open the valve. Adjusting the preset comparator voltage levels results in a valve timing change. This variable control can be accomplished while the unit is in operation. It should be noted there that the intake valve timing as well as stroke can be varied in this manner.

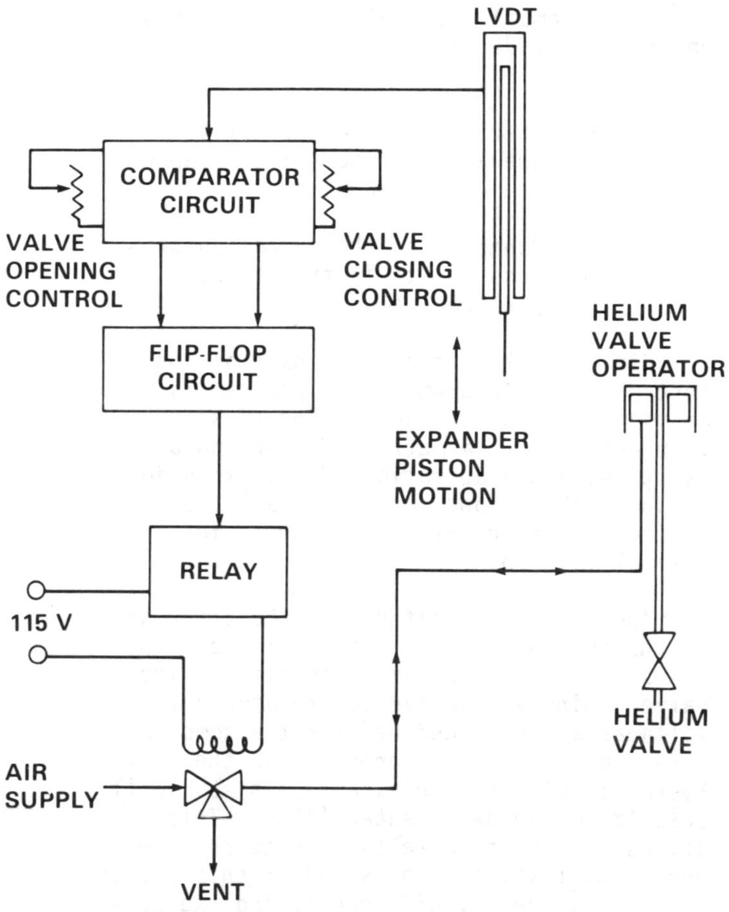

Figure 4. Block diagram of electronic valve timing control.

VARIABLE GEOMETRY

Helium liquefiers have operated successfully for a number of years with fixed stroke, fixed valve timing and variable speed. It has been observed through extensive Navy liquefier testing that these machines do not operate at their design point for maximum cooling except during certain steady state operating conditions. Additionally, an experienced operator is required to cool the machine down and maintain liquid helium production.

In adapting superconductive electric drive machinery to a Navy ship, it becomes necessary for the helium liquefication system to operate efficiently and automatically in several operational modes.

First, the liquefier must have the capability of cooling down itself and/or a large (4,500 kg) magnet system in very short period of time (72 hours). The liquefier must operate at or very near optimum operating conditions over a wide range of temperatures. This can best be accomplished by varying the valve timing, stroke and speed as required by the operating temperatures to keep the liquefier operating close to its design points.

A second mode of operation requires the maintenance of optimum operation at off design conditions due to degradation of performance of various components, and as a result of different refrigeration loads. These changes may be small in magnitude, but are very critical in providing a long term, reliable system for superconductive magnets or cryogenic sensors.

A small digital computer with appropriate software and a feedback loop could provide the automatic control of a variable geometry helium expansion engine in such a helium liquefier.

A DEC PDP 11/30 computer was selected to be used in these experiments. It contains 128K memory. The software package included the DEC Laboratory Program Accessory (LPA) package, which provides the clock functions and the analog to digital and digital to analog sweeps. The control program was written in FORTRAN IV. While this high level language uses much more computer time and space for execution of statements than the equivalent in assembly or machine language, it was chosen for its simplicity of writing, editing, and modifying the program. It should be pointed out here that an operational shipboard system would be programmed in a more basic language or even hard-wired in a micro processor of some type.

COMPUTER CONTROL LOGIC

The control logic for the automatic computer operation was developed through a series of experiments. These experiments required an operator to manually set the reference voltages of the comparators by adjusting a set of potentiometers as a function of the expansion engine temperature. This information was then incorporated in a fortran program. A set of helium temperatures is read into the computer through analog-to-digital (A/D) converters. Timing voltages were determined by the computer program and transfered to the comparator through digital-to-analog (D/A) converters. Figure 5 indicates a block diagram of the logic sequence for starting and operating the expansion engine with the computer.

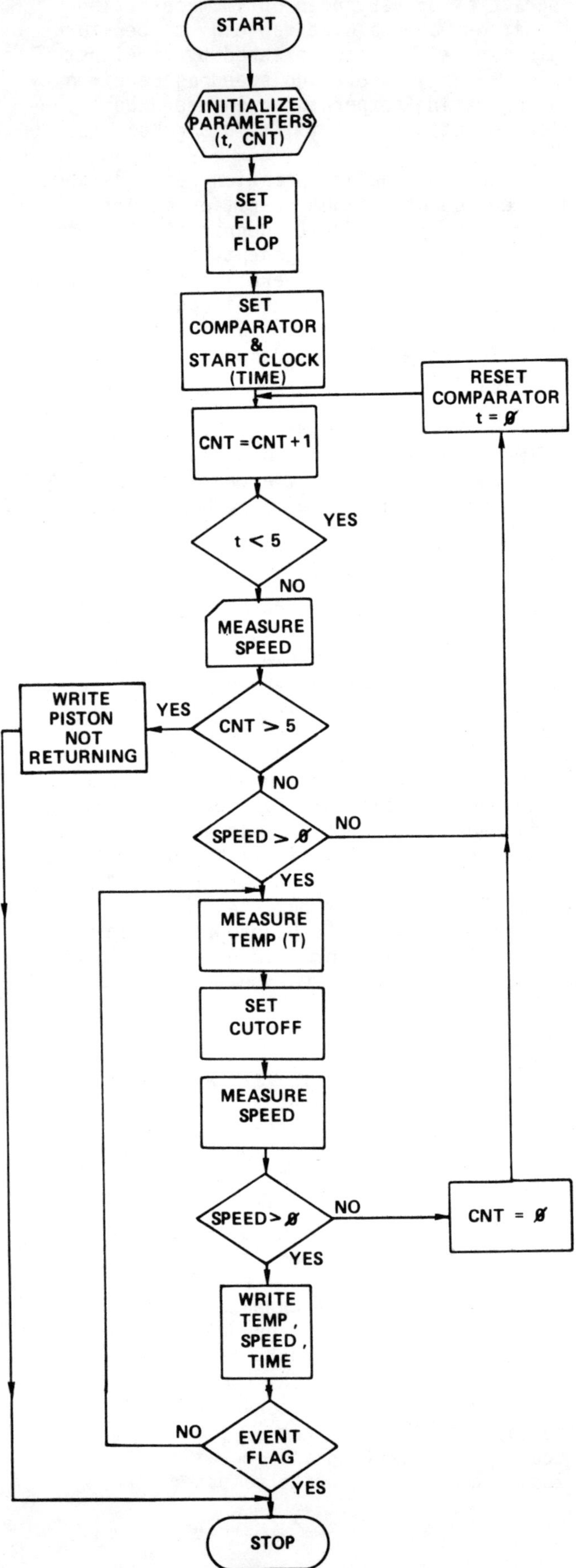

Figure 5. Computer control flow chart.

The automatic control of the expansion engine is divided into two operating features:

1. Start up maintenance of the operational mode, i.e. keeping the engine running.

2. Controlling the expansion process to attain the desired performance.

Figure 5 shows the flow chart of the control logic program. This program is divided into two sections. The first section establishes a method of starting up and maintaining the operation of the expansion engine. The second section establishes a procedure for operating the engine in some optimal fashion.

The logic associated with the start up of the engine requires set-up of the flip flops in the correct order to start. The valve timing and active accumulator reference voltages are next defined for the comparators. Because of the pressure on the hydraulic piston, the expansion piston will default to top dead center (TDC). This allows the inlet valve to operate and the engine will start. The speed of this piston can be measured by differentiating the output of the LVDT and thus determine if the engine started. If for some reason it did not start, the valve timing can be adjusted and restarted. If several attempts fail to restart the engine, one may assume that the piston is not returning to TDC, and the flip flops are not being reset. Possible reason for this to occurr might be contamination or malfunction of the expansion engine.

The second section of the computer control scheme adjusts the valve timing with respect to piston position based on temperature and pressure. These adjustments are made on the valve timing to give the optimum performance of the engine. In addition, the speed of the engine is checked to periodically verify the expansion engine is operating. A continuous update of the status of the engine can be displayed.

The hardware necessary to control the speed and stroke of the expander are complete. The algorithm to control speed and stroke as a function of required helium mass flow is not yet complete.

Future work includes expansion of the program to control the speed and valve timing of two expansion engines simultaneously. This program will be used to control the expansion engines in an existing helium liquefier. Software to control the Joule-Thompson (J-T) valve will also be written for this liquefier.

CONCLUSIONS

The operation and control of a single expansion engine using a digital computer was performed as a part of a longer term development of a fully automatic control for helium liquefiers. The system was started at room temperature, 300K, and cooldown to a minimum of 35K with one stage of expansion. The valve timing was varied according to a predetermined function of temperature to achieve maximum refrigeration throughout the cooldown cycle. Implementation of the control system to vary the valve timing and maintain continuous operation of this hydraulically controlled expansion engine was demonstrated.

These results established the feasibility of using a computer control system to operate an expansion engine automatically at various operating conditions. The best performance of an expansion engine can only be obtained by varying the valve timing, speed, and stroke of the engine as a function of the temperature flow, and pressure of the expanding gas. The optimization of the cooling capacity of the expansion engine can be regulated based on these parameters by the use of an automatic control. Using a computer control system with a hydraulically operated expansion engine has provided this required flexibility which maximizes the cooling ability of the expansion engine with little or no operation assistance.

LITERATURE CITED

1. G. Patton, G. Green, K. Dunn, and V. Dilling, in "Advances in Cryogenic Engineering, Vol. 27," Plenum Press, New York (1982), p. 641.

PREDICTION OF THE VISCOSITY OF PURE AND MIXED CRYOGENIC FLUIDS

Viscosity and thermal conductivity play an important role in engineering design. Recently, a predictive corresponding states model for these properties in nonpolar mixtures has been developed. The method, which is applicable to the entire range of fluid states, does not require any transport data in the predictions. This manuscript summarizes recent studies dealing with the applicability of a corresponding states principle for fluid viscosity. Comparisons of predictions and experiment for eight pure cryogenic fluids and three mixtures are presented.

JAMES F. ELY
and
JOLENE K. BAKER
Chemical Engineering Science Division
National Engineering Laboratory
National Bureau of Standards
Boulder, Colorado 80303

Transport properties of fluid mixtures play a key role in engineering design. Even though they are generally not as important as equilibrium properties such as enthalpies and phase equilibria, uncertainties of 10-20 percent can affect compressor and pump sizes and surface area requirements in heat exchangers. To illustrate this point, Mickley and Hanley (1) have recently shown that uncertainties of 10 percent in the heat capacity viscosity and thermal conductivity can give rise to oversizing of heat exchangers by 10-20 percent.

During the past two years a model for the prediction of the viscosity and thermal conductivity of nonpolar pure fluids and their mixtures has been developed (2-4). This model, which is based on a one-fluid conformal solution principle, used a methane reference fluid with appropriate (large) corrections for noncorrespondence in the above mentioned transport properties. The model does not require any transport data for the calculation. Although the model was quite successful, the use of empirical corrections for noncorrespondence was unsatisfactory. In addition methane, which has a relatively high reduced triple point, was unsuitable for corresponding states calculations for higher molecular weight materials. Methane was chosen as the reference fluid since at that time it was the only fluid for which there were accurate wide range correlations for the pVT, viscosity and thermal conductivity.

Recently, we have obtained wide range equations of state and transport property surface correlations for propane (5). Propane, which has the lowest reduced triple point of the paraffin homologous series, is an ideal reference fluid. In addition, one would expect the transport properties of heavier hydrocarbons to correspond better with propane than methane, owing to the fact that first members of the homologous series are notoriously anomalous in their physical property behavior.

In this manuscript we report the results of applying the propane based corresponding states viscosity model to several cryogenic fluids and their mixtures. The emphasis of this study was to investigate the degree to which fluid viscosity obeys this corresponding states principle. Results are presented for eight pure fluids and three mixtures.

MODEL DESCRIPTION

The viscosity model is based on four fundamental assumptions: 1) the viscosity of

Paper is a contribution from the U. S. National Bureau of Standards, mixture or pure fluid can be equated to that of a hypothetical pure fluid; 2) the hypothetical pure fluid viscosity may be evaluated via a corresponding states principle; 3) the reference fluid state point used in the transport calculation may be evaluated via equilibrium corresponding states; and 4) mixture size difference effects may be taken into account with the Enskog hard sphere theory. Mathematically we have

$$\eta_{mix}(\rho,T,\{x_i\}) = \eta_o(\rho_o,T_o)\,F_\eta + \Delta\eta_{mix}^{ENS}$$

where η is the viscosity, ρ is the density, T is the absolute temperature $\{x_i\}$ denotes the mixture composition, F_η is a dimensional factor and $\Delta\eta_{mix}^{ENS}$ is a correction for size differences given in (3). The subscript o denotes a reference fluid value. The reference fluid density and temperature are defined by

$$\rho_o = \rho h_x \quad \text{and} \quad T_o = T/f_x \qquad (1)$$

where f_x and h_x are given in terms of the critical constants and shape factors of Leland and Leach (6). For a pure fluid

$$f_x = (T_c^x/T_c^o)\,\theta(T_r,V_r,\omega)$$

and

$$h_x = (V_c^x/V_c^o)\,\phi(T_r,V_r,\omega)$$

The subscript r denotes a value reduced by the critical point and ω is Pitzer's acentric factor. For mixtures, van der Waals' one-fluid mixing rules are used:

$$f_x h_x = \sum\sum x_i x_j \; f_{ij} h_{ij}$$

and

$$h_x = \sum\sum x_i x_j \; h_{ij}$$

The dimensional factor for viscosity is given by

$$F_\eta = \left(\frac{M_x}{M_o}\right)^{1/2} f_x^{1/2}\, h_x^{-2/3}$$

where for the mixture viscosity, the effective mass is defined by

$$h_x^{4/3} f_x^{1/2} M_x^{1/2} = \sum\sum x_i x_j \, h_{ij}^{4/3} f_{ij}^{1/2} M_{ij}^{1/2} .$$

In these equations $f_{ij} = (f_i f_j)^{1/2}$, $h_{ij} = \frac{1}{8}(h_i^{1/3} + h_j^{1/3})$, $M_{ij} = 2M_iM_j/(M_i + M_j)$, and x_i is the mole fraction of component i in the mixture.

SHAPE FACTOR CALCULATIONS

In order to avoid any errors associated with the shape factors and predicted densities, the shape factors were determined exactly from equations of state for the pure fluids. The equations of state incorporated in this calculation were of the Stewart-Jacobsen form (7) and were fit to both saturation and single phase data.

Given the equations of state, the shape factors were determined by simultaneous solution of the following equations.

$$\frac{A^r(\rho,T)}{RT} = \frac{A_o^r(\rho_o,T_o)}{RT_o}$$

and

$$Z(\rho,T) = Z_o(\rho_o,T_o)$$

where A^r is a residual Helmoltz free energy relative to the ideal gas at the same temperature and density and Z is a compressibility factor. The quantities ρ_o and T_o are defined in Equation (1).

Even though there is no uncertainty in the densities and shape factors obtained for the pure fluids, some error in the predicted mixture density may be introduced by the one-fluid model as described previously. Recent studies have shown that these errors are usually less than one percent for the fluids considered in this study.

RESULTS AND DISCUSSION

Figures 1-3 show the observed differences between the calculated and experimental viscosities of several cryogenic fluids and mixtures and Table 1 summarizes the results. As one can see, there are no uniform trends to the deviation patterns. The predictions for n-butane, nitrogen, carbon dioxide and ethylene are very good while deviations for the other pure fluids show systematic trends which cannot be rationalized on the basis of molecular structure. It is interesting to note that the systematic deviations mentioned above seem to start at a reduced density of two where the slope of the viscosity versus density curve becomes very steep.

Based on the results depicted in Figures 1-3 one can make several observations and conjectures. First, one can conclude that the viscosity does obey a corresponding states principle below densities of twice the critical density. Second, there appears to be a pseudo-phase transition at $2\rho_c$ which may be conjectured to be due to a substantial reduction in the kinetic mean free path of the molecules. It is also observed that in the case of i-butane there is a substantial increase in the fluid viscosity over that of n-butane which one may attribute to molecular branching. Finally, one may postulate the need for a mass shape factor in a viscosity model. This mass shape factor, which may involve the moments of inertia of the molecules, may enable one to account for structural differences in the molecules.

ACKNOWLEDGMENTS

The authors would like to acknowledge the support of the NASA-Lewis Research Center and the Office of Standard Reference Data for their support of this work. Ms. Karen Bowie helped substantially in the preparation of the manuscript.

LITERATURE CITED

1. Mickley, M. and Hanley, H. J. M., Proc. 1982 ASME Winter Annual Meeting, Paper 82-WA/HT-60.

2. Ely, J. F. and Hanley, H. J. M., Ind. Eng. Chem. Fundam. 20, 323 (1981).

3. Ely, J. F., J. Res. Nat. Bur. Stand. (U.S.) 86, 597 (1981).

4. Ely, J. F. and Hanley, H. J. M., Ind. Eng. Chem. Fundam. 22, 90 (1983).

5. Ely, J. F., Proc. 61st Conv. Gas Processors Assn., p 9 (1981).

6. Leach, J. W., Chappelear, P. S. and Leland, T. W., AIChE J. 14, 568 (1968).

7. Jacobsen, R. T. and Stewart, R. B., J. Phys. Chem. Ref. Data 2, 757 (1973).

Table 1. Comparison of Calculated and Predicted Viscosities for Selected Cryogenic Fluids

Fluid	N	AAD, %	BIAS, %	RMS, %
Methane	218	8.37	7.48	5.94
Ethane	658	8.34	1.04	11.71
n-Butane	282	2.00	1.21	2.45
i-Butane	120	7.92	-6.14	10.17
Nitrogen	81	3.96	-3.75	2.46
Argon	368	6.59	-5.85	7.97
CO_2	1320	5.85	2.04	9.99
Ethylene	247	4.26	.33	5.51
Methane/Nitrogen	306	7.57	5.13	7.24
Methane/Ethane	319	9.50	4.19	15.45
Methane/Propane	134	4.24	2.09	5.12
Overall	4748	5.87	1.10	6.31

ADD = average absolute deviation; BIAS = average deviationν RMS = root mean square deviation.

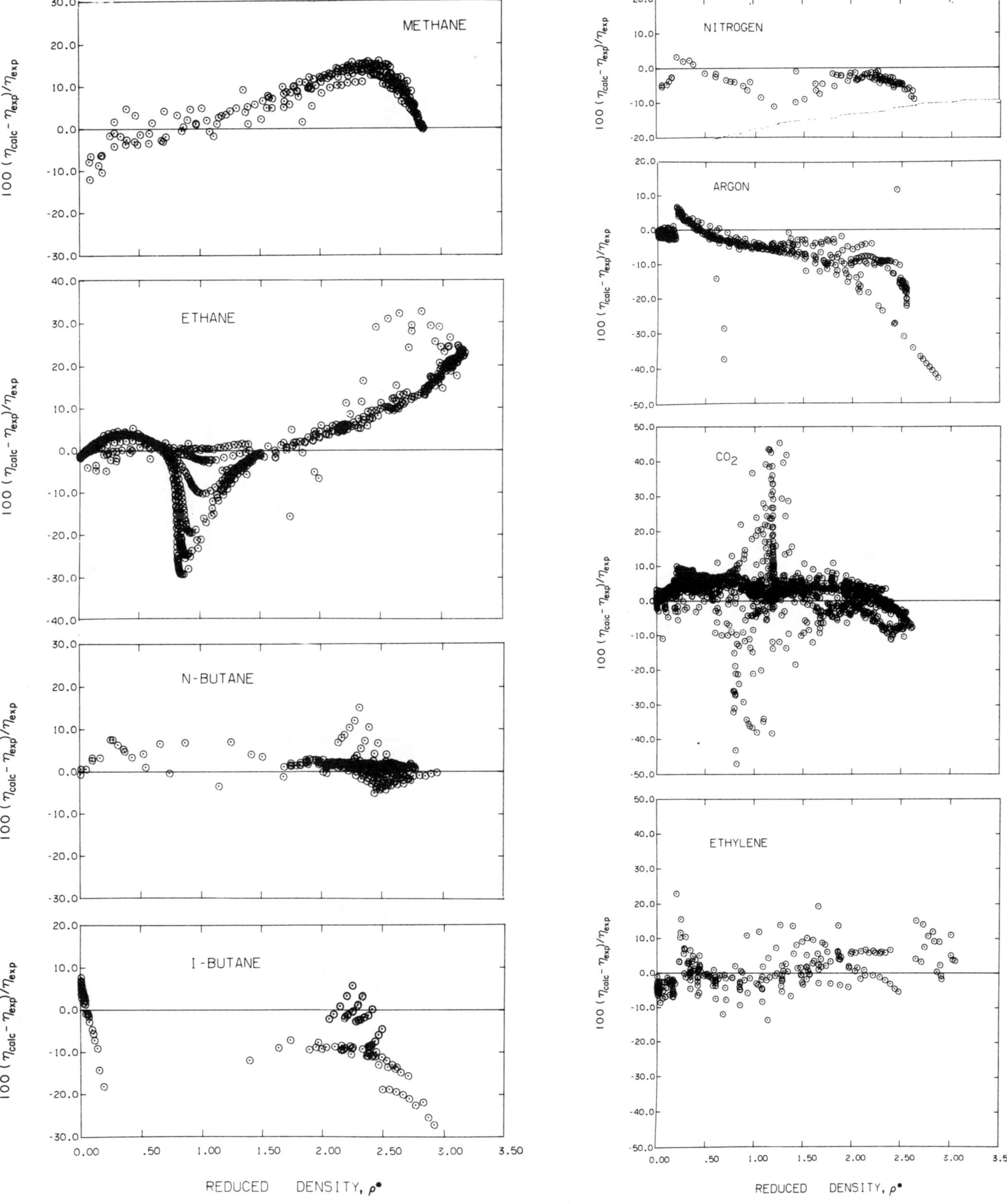

Figure 1. Comparison of predicted and experimental viscosities of methane, ethane, n-butane and i-butane.

Figure 2. Comparison of predicted and experimental viscosities of nitrogen, argon, carbon dioxide, and ethylene.

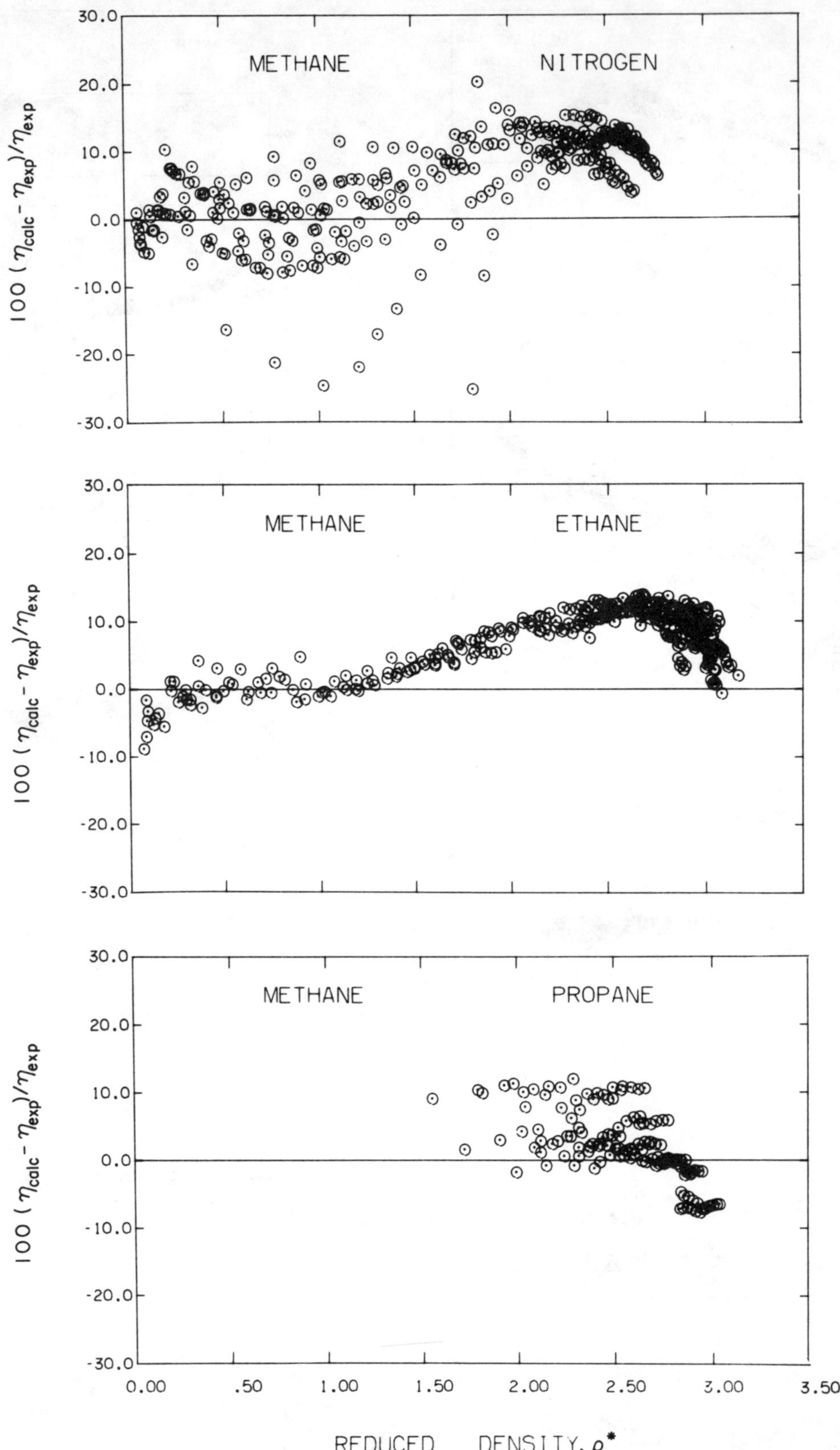

Figure 3. Comparison of predicted and experimental viscosities of methane/nitrogen, methane/ethane and methane/propane binary mixtures.

THE STATUS OF THERMOPHYSICAL PROPERTIES DATA FOR PURE FLUIDS AND MIXTURES AT LOW TEMPERATURES

This paper discusses the need for and the availability of both thermodynamic and physical properties data for the fluids generally encountered in low-temperature processing. The important gaps in the existing data are noted and recommendations are made for future experimental programs.

A. J. KIDNAY
N. A. OLIEN
M. J. HIZA

The objective of this paper is to address the following four questions as they pertain to cryogenic processing:

- Why are data on thermophysical properties important?
- What data are needed?
- What data are available?
- What work remains to be done?

THE IMPORTANCE OF THERMOPHYSICAL PROPERTIES DATA

Surprisingly few studies have been carried out on the importance of properties data in chemical engineering. The work of Zudkevitch (1), Zudkevitch and Gray (2), and Chappelear, Chen, and Elliott (3) stand out as the principal studies on the impact of inaccurate data on both plant performance and economics. The latter authors studied the effect of different enthalpy, entropy, and K-value correlations on the performance of two typical turboexpander plants. They concluded that variation in predicted yearly income from the two plants due solely to the different values predicted by the various correlations tested amounted to $210,000 for one plant and $237,000 for the other. The original papers should be consulted for further comparisons and the specifics of the calculations.

DATA NEEDS FOR CRYOGENIC PROCESSING

To determine the type and relative importance of the thermophysical properties data used in cryogenics reference will again be made to the work of Zudkevitch (1, 2). Table 1 is an abridged and modified version of tables in his work.

The following brief and simplified discussion of how the various types of data may be useful in process calculations will hopefully lend support to the contention that accurate data are essential for good process performance.

Phase Equilibria in Binary and Multicomponent Systems

The equilibria of principal interest are liquid-vapor, solid-vapor, liquid-liquid (or liquid-liquid-vapor), and solid-liquid (or solid-liquid-vapor).

Vapor-liquid equilibria are essential in the design of separation units, notably single pass condensers or vaporizers, and distillation columns. Since it is impossible to gather data for all conceivable operating conditions, the main use of experimental measurements is to evaluate interaction coefficients in equations of state and to determine liquid phase activity coefficients. The values thus obtained are used for extrapolation or interpolation into regions where no data exist. For example, the equations used to calculate VLE are

$$k_i \equiv y_i/x_i = \phi_i^{l} / \phi_i^{v} \qquad (1)$$

or

$$\frac{\gamma_i P_i^{sat} \phi_i^{sat} \quad \exp\left[v_i^l (P-P_i^{sat})/RT\right]}{\phi_i^v P} \quad (2)$$

where y_i and x_i are the vapor and liquid mole fractions, ϕ_i^l and ϕ_i^v are the fugacity coefficients of component i in the liquid and vapor phases, γ_i is the liquid phase activity coefficient, ϕ_i^{sat} is the fugacity coefficient of pure component i at its saturation pressure, v_i^l is the liquid molar volume of the pure component, P_i^{sat} is the vapor pressure of the pure component, and P, R, and T are the pressure, gas constant, and temperature. As mentioned, in order to calculate either ϕ_i^l or ϕ_i^v an empirical binary interaction constant, usually given the symbol k_{ij}, must be used in conjunction with a suitable equation of state (4). The k_{ij} is normally considered to be a function of only the molecular species, and to be independent of composition, temperature, and pressure. Thus, in principle, only a single VLE measurement would suffice for the evaluation of k_{ij}. However, in practice, data covering an extensive range of conditions are used since generally k_{ij} is not a true constant. In extending the calculations to multicomponent systems, it is assumed that only binary k_{ij} parameters are required. If instead of ϕ_i^l, the liquid phase activity coefficient, γ_i, is desired, then extensive VLE data are necessary, since γ_i depends on composition, temperature, and, to a limited extent, the pressure.

Information on solid-vapor equilibria is obviously essential in those systems where solid formation would result in the fouling or plugging of heat exchangers or process lines. From the process engineer's viewpoint, the goal is to determine what operating conditions to avoid. In solid-vapor phase equilibria, the principal relation is

$$\frac{y_i P}{P_i^{sat}} = \frac{\phi_i^{sat}}{\phi_2^v} \exp\left[v_i^s (P-P_i^{sat})/RT\right] \quad (3)$$

where v_i^s is the molar volume of the solid and the remaining notation is the same as in equations (1) and (2). In the development of this equation, the solid phase is assumed to be pure. As in the case of VLE, the major use of the phase equilibria data is in evaluating the binary interaction constant, k_{ij}, used in the equation of state calculation for ϕ_i^v.

Liquid-liquid equilibria are much less important (and much less well understood) than vapor-liquid equilibria, but there a few systems, notably He^3-He^4, that are of considerable interest. The equilibrium relationship is

$$x_i^\alpha \gamma_i^\alpha = x_i^\beta \gamma_i^\beta \quad (4)$$

where the α and β refer to the two liquid phases. As mentioned previously, extensive phase equilibria data are necessary for the correlation of the activity coefficients.

As with solid-vapor data, the principal use of solid-liquid equilibria data is to determine what operating conditions to avoid, so that solid formation will not occur. The equilibrium relationship is (approximately)

$$x_i = \frac{P_i^{sat}}{\gamma_i P_{i,scl}^{sat}} \quad (5)$$

where P_i^{sat} is the vapor pressure of the pure solid and $P_{i,scl}^{sat}$ is the vapor pressure of the pure subcooled liquid. Again, extensive data are required for proper evaluation of the activity coefficient.

(Vapor Pressure Including Critical Properties)

When dealing with pure fluids, the need for vapor pressure data is obvious, but the importance of pure fluid properties in mixture calculations is often overlooked. Examination of equations (2), (3), and (5) shows that the correlation or prediction of vapor-liquid, solid-vapor, and solid-liquid equilibria requires accurate vapor pressure data for all of the subcritical components of the mixture.

Since the Principle of Corresponding States is widely used in chemical engineering calculations, the need for accurate critical properties is again obvious.

Pressure-Density-Temperature (Pure Fluids and Mixtures)

Where pure fluids are concerned, the usefulness of volumetric data needs no elaboration, but as in the case of vapor pressures, the value of this type data when working with mixtures is often overlooked. Any mixture predictive or correlative technique must rely heavily on pure fluid properties in order to have any hope of being generally applicable over wide ranges of components, composition, temperature, and pressure. For example, pure fluid molar volumes are needed in high pressure phase equilibria calculations to evaluate the effect of total system pressure on the pure fluid fugacity (the Poynting correction, $\exp\left[v_i(P-P_i^{sat})/RT\right]$) and partial molar volumes are required to

TABLE 1
The Relative Importance of Thermophysical Properties Data in Cryogenic Processing

Operation or Area	Type of Data: Vapor Pressure (including critical properties)	Phase Equilibria in Multicomponent Systems	P-ρ-T (pure fluids and mixtures)	Calorimetric Properties (pure fluids and mixtures)
Distillation	A	A	B	B
Compression & Expansion	A	A	A	A
Heat Exchange	B	A	C	A
Metering and Sales	B	B	A	B
Storage	B	A	B	A
Safety	B	A	B	A

A = most important
B = very important
C = minor importance

evaluate the effect of pressure on the liquid phase activity coefficient (4, p 188-9). In the area of metering and sales, highly accurate volumetric data on mixtures are obviously essential for custody transfer.

Calorimetric Properties (Pure Fluids and Mixtures)

Thermal properties (enthalpy and heat capacity) are required for proper design and performance evaluation in heat exchangers, compressors, pumps, and expanders. For example, work requirements in an adequate flow process are given by $w_s = -\Delta h$ where Δh is the enthalpy change of the process fluid as it passes through the compressor. Calorimetric data are also useful in phase equilibria, an example being the use of the heat of mixing (or excess enthalpy) to determine the temperature effect on the excess Gibbs energy, and thus the activity coefficient, through the relations

$$\frac{\partial(g^e/T)}{\partial T} = \frac{-h^e}{T^2} \qquad (6)$$

and

$$RT \ln \gamma_i = \left(\frac{\partial g^e}{\partial n_i} \right)_{P,T,n_j} \qquad (7)$$

In these equations, g^e and h^e are the molar excess Gibbs energy and the molar excess enthalpy (or heat of mixing).

Calorimetric properties of pure fluids and mixtures are important in themselves, but they also serve as a powerful tool in testing the accuracy of predictive techniques.

AVAILABILITY OF THERMOPHYSICAL PROPERTIES DATA

The review will be divided into two major sections - pure fluids and mixtures.

Pure Fluid Properties

Accurate knowledge of the thermophysical properties of pure fluids is important for both custody transfer and design of processes. Examples are hydrogen and oxygen in the space program, oxygen and nitrogen as industrial and medical gases, ethylene as a chemical feedstock, etc. More and more, however, these pure fluids take on importance as constituents of mixtures and the major portion of this review will address that aspect.

The properties we are mostly concerned with for the pure fluids are: PVT or volumetric properties; vapor pressure; orthobaric densities; heat capacity or enthalpy data; sound speed and the transport properties, viscosity and thermal conductivity.

Table 2 lists the primary sources of standard data and wide range correlations for the thermodynamic properties of the pure fluids considered in this review. Table 3 lists sources of data for the transport properties of pure fluids. Generally speaking, experimental transport data of high accuracy are difficult to locate.

TABLE 2

THERMODYNAMIC PROPERTIES OF PURE FLUIDS

H_2	McCarty, R.D., Hord, J., Roder, H.M., Nat. Bur. Stand. Monograph 168 (1981)
He	McCarty, R.D., J. Phys. Chem. Ref. Data 2, 923 (1973) and Angus, S., et al., Intern. Thermodynamic Tables of the Fluid State - 4, Butterworths (1977)
Ne	Jacobsen, R.T., Stewart, R.B., Teng, J.C.J., Adv. Cryogenic Eng. 27, 911, Plenum Press (1982)
N_2	Younglove, B.A., J. Phys. Chem. Ref. Data 11, Suppl. 1, 329 pp (1982)
Kr	Juza, J., and Sifner, O., Acta Tech. CSAV (Prague) 21, 1-32 (1976)
Xe	Juza, J., and Sifner, O., Acta Tech. CSAV (Prague) 22, 1-32 (1977)
Ar	Younglove, B.A., ibid.
O_2	Younglove, B.A., ibid.
F_2	Prydz, R., and Straty, G.C., Nat. Bur. Stand. Tech Note 392 revised (1973)
Cl_2	Vargaftik, N.B., "Tables on the Thermophysical Properties of Liquids and Gases", J. Wiley and Sons (1975)
CO	Din, F., Thermodynamic Functions of Gases, Vol I, Butterworths (1956) and Goodwin, R.D., J. Phys. Chem. Ref. Data 12 (1983) to be published.
CO_2	Angus, S., et al., Intern. Thermodynamic Tables of the Fluid State - 3, Pergamon (1976)
NO	No comprehensive data source available
N_2O	No comprehensive data source available
SO_2	Vargaftik, N.B., ibid.
NH_3	Haar, L., and Gallagher, J.S., J. Phys. Chem. Ref. Data 7, 635 (1978)
NF_3	Younglove, B.A., ibid.
CH_4	Goodwin, R.D., Nat. Bur. Stand. Tech. Note 653 (1974) and Angus, S., et al., Intern. Thermodynamic Tables of the Fluid State - 5, Pergamon (1978)
C_2H_6	Goodwin, R.D., et al., Nat. Bur. Stand. Tech. Note 684 (1976)
C_3H_8	Goodwin, R.D., and Haynes, W.M., Nat. Bur. Stand. Monograph 170 (1982)
n-C_4H_{10}	Haynes, W.M., and Goodwin, R.D., Nat. Bur. Stand. Monograph 169 (1982)
i-C_4H_{10}	Goodwin, and Haynes, W.M., Nat. Bur. Stand. Tech. Note 1051 (1982)
C_2H_4	Younglove, B.A., ibid.
C_3H_6	Angus, S., et al., Intern. Thermodynamic Tables of the Fluid State - 7, Pergamon (1980)
Refrigerants	
	ASHRAE Handbook 1981 Fundamentals, Amer. Soc. Heating, Refrig. Air Cond. Eng. (1981)

TABLE 3

TRANSPORT PROPERTIES OF PURE FLUIDS

H_2	McCarty, R.D., Hord, J., Roder, H.M., Nat. Bur. Stand. Monograph 168 (1981)
He	Vargaftik, N.B., "Tables on the Thermophysical Properties of Liquids and Gases", J. Wiley & Sons (1975)
Ne	Vargaftik, N.B., ibid.
Ar	Younglove, B.A., J. Phys. Chem. Ref. Data 11, Suppl. 1, 329 pp (1982)
Kr	Hanley, H.J.M., McCarty, R.D., Haynes, W.M., J. Phys. Chem. Ref. Data 3, 979 (1974)
Xe	Hanley, H.J.M., ibid.
N_2	Younglove, B.A., ibid.
O_2	Younglove, B.A., ibid.
F_2	Haynes, W.M., Physica 76, 1 (1974)
Cl_2	Vargaftik, N.B., ibid.
CO	Vargaftik, N.B., ibid.
CO_2	Vargaftik, N.B., ibid.
NO	Barua, A.K., et al., Int. J. Heat Mass Transfer 12, 587 (1969)
N_2O	Hanley, H.J.M., Nat. Bur. Stand. Tech. Note 693 (1977)
SO_2	Vargaftik, N.B., ibid.
NH_3	Vargaftik, N.B., ibid.
NF_3	Vargaftik, N.B., ibid.
CH_4	Hanley, H.J.M., Haynes, W.M., McCarty, R.D., J. Phys. Chem. Ref. Data 6, 597 (1977)
C_2H_6	Hanley, H.J.M., Gubbins, K.E., Murad, S., J. Phys. Chem. Ref. Data 6, 1167 (1977)
C_3H_8	Holland, P.M., Hanley, H.J.M., Gubbins, K.E., Haile, J.M., J. Phys. Chem. Ref. Data 8, 559 (1979)
n-C_4H_{10}	Ely, J.F., and Hanley, H.J.M., Nat. Bur. Stand. Tech. Note 1039 (1981)
i-C_4H_{10}	Ely, J.F., and Hanley, H.J.M., ibid.
C_2H_4	Hanley, H.J.M., J. Phys. Chem. Ref. Data 12, (1983) to be published.
C_3H_6	Vargaftik, N.B., ibid.
Refrigerants	
	ASHRAE Handbook 1981 Fundamentals, Amer. Soc. Heating, Refrig. Air Cond. Eng. (1981)

Mixture Properties

The equilibria to be considered are solid-vapor, solid-liquid (solid-liquid-vapor), and liquid-vapor equilibria as well as volumetric properties for both liquid and vapor mixtures, calorimetric measurements, and transport properties.

It is obviously impossible, in a short review paper such as this one, to list the literature references for all cryogenic fluids and mixtures, and thus we must limit ourselves to a general discussion and refer the reader to published bibliographies for the specific references. For the mixture data, the source is the recent bibliography of Hiza, Kidnay, and Miller (5). Additional useful bibliographies containing cryogenic thermophysical properties are those of Wichterle, Linek, and Hāla (6), and Wisniak and Tamir (7).

With regard to the transport properties of mixtures, the availability of accurate, wide-range experimental data is extremely limited. Some progress is being made for selected binary systems by Kestin (8) and Diller (9). The most useful source of mixture transport property data is the predictive technique of Ely and Hanley called TRAPP (10, 11, 12).

In Figures 1 through 6, the availability of binary data for systems of interest in low temperature processing is summarized. The pressure and temperature regions covered by the data and the specific references are given by Hiza, Kidnay, and Miller (5). The calorimetric properties of Figure 6 generally consist of heats of mixing, heat capacities, and Joule-Thomson coefficients.

DATA NEEDS FOR THE FUTURE

In what follows the reader should be aware that the discussion of future data needs is not based on a critical evaluation of all available data, but consists of the subjective judgements of the authors based on their personal experience. Since the needs of individuals and groups may vary substantially, there will be undoubtedly some disagreement with the selections made in this paper.

Pure Fluids

Table 4 gives specific recommendations for pure fluid measurements which, in our opinion, have a high priority for the use of the fluids as cryogens and as constituents of mixtures ranging from liquefied natural gas to coal conversion products. Table 5 summarizes the general data needs for pure fluids.

Mixtures

In assessing the needs for future experimental work, we will be considerably more selective than in presenting the availability diagrams of the previous section. Specifically, the discussion will be limited to systems of substantial engineering or scientific interest consisting of combinations of H_2, He^4, Ne, N_2, O_2, Ar, CO_2, CH_4, C_2H_6, C_8H_8, n-C_4H_{10}, iso-C_4H_{10}, and H_2O.

The authors' judgements on the current state of affairs are summarized in Figures 1 through 6.

Solid-vapor equilibria

The most interesting systems for which no data exist are:

1) H_2 with the heavier hydrocarbons C_3H_8, and C_4H_{10},
2) He with C_2H_6 and heavier hydrocarbons,
3) O_2 with CO_2, and
4) the primary air components (N_2, O_2, Ar) with n-C_4H_{10}.

The solid-vapor phase data for systems containing water are very limited and additional work would be desirable.

Solid-liquid (solid-liquid-vapor) equilibria

The H_2, He, and Ne systems appear to be the most deficient in data. A comprehensive study of the low molecular weight systems would be very interesting. Several of the N_2 and CH_4 systems have very limited data available and additional research seems justified.

Liquid-liquid equilibria

Since it is not always obvious which systems will exhibit these equilibria, it is unrealistic to make recommendations for systems that have not been previously studied.

Liquid-vapor equilibria

Figure 4 shows that, as would be expected, VLE data are the most abundant type of phase equilibria. There are over 1,000 references to binary systems available, covering wide ranges of temperatures and pressures. The principal deficiencies are shown in the figure, but it is certain that there are many systems where critical data evaluation would show that many of the existing measurements are of low quality and should be discarded.

Volumetric and calorimetric equilibria

The type of measurements covered in these categories varies substantially and thus it is quite difficult to make specific recommendations. Some of the more pronounced deficiencies are indicated in Figures 5, 6, and 7.

TABLE 4
PROPERTIES OF FLUIDS - SPECIFIC DATA NEEDS

H_2	Transport Properties, 20-300K
He	PVT, 2-20K; Transport Properties, 2-300K
Ne	Orthobaric Densities; Low Temperature Derived Property Measurments
Ar, Kr, Xe	New Critical Evaluation of Entire PVT Surface
N_2	High Pressure Heat Capacity Measurements
O_2	PVT above 350K; C_v below 250K
Cl_2	Comprehensive PVT and Transport Property Measurements

CO	Critical Parameter and Orthobaric Density Measurements
CO_2	C_v and Sound Speed Measurements in Liquid Region
NO, N_2O, SO_2	Comprehensive PVT and Trans port Property Measurements
NF_3	Transport Property Measurements
CH_4, C_2H_6	Low Density PVT Measurements, near Critical and above
i-C_4H_{10}	PVT and Orthobaric Densities, 300-408K
i-C_4H_{10}, n-C_4H_{10}	Transport Property Measurements in General and Dilute Gas Thermal Conductivity

TABLE 5
PROPERTIES OF FLUIDS -
GENERAL DATA NEEDS

- Accurate Measurements of Derived Properties (Heat Capacity, Sound Speed) over Substantial Density Ranges to Test and Optimize Equations of State
- Measurements of Saturated Vapor Densities for Nearly All Pure Fluids
- High Temperature PVT Measurements of Nearly All Fluids, but Especially C_4's and above
- Systematic Evaluation and Correlation of Data for Alkanes C_5 and Above
- Measurements of Self-diffusion Coefficients for Nearly All Fluids

Literature Cited

1. Zudkevitch, D.; Hydrocarbon Process., March 1975, pp. 97-102.

2. Zudkevitch, D., and Gray, R.D., Jr.; Adv. in Cryogenic Engineering, Vol. 20 (K.D. TImmerhaus, ed.) Plenum Press, NY (1975) pp. 102-123.

3. Chappelear, P.S., Chen, R.J.J., and Elliot, D.G.; Hydrocarbon Process., September 1977, pp. 215-217.

4. Prausnitz, J.M.; "Molecular Thermodynamics of Fluid-Phase Equilibria", Prentice Hall, Englewood Cliffs, NJ (1969).

5. Hiza, M.J., Kidnay, A.J., and Miller, R.C.; Equilibrium Properties of Fluid Mixtures: A Bibliography of Experimental Data on Selected Fluids, IFI/Plenum, New York (1982), 246 pages.

6. Wichterle, I., Linek, J., and Hāla, E.; Vapor-Liquid Equilibrium Data Bibliography, Elsevier, Amsterdam (1973) 1053 pages: Supplement I, Elsevier, Amsterdam (1976) 333 pages: Supplement II, Elsevier, Amsterdam, (1979) 286 pages.

7. Wisniak, J., and Tamir, A.; "Mixing and Excess Thermodynamic Properties", Elsevier, Amsterdam (1978) 935 pages.

8. Kestin, J. Nagasaka, Y. Wakeham, W.A.; Physica 113A, 1 (1982).

   Fleeter, R., Kestin, J., Nagasaka, Y., Shankland, I.R., Wakeham, W.A.; Physica 111A, 404 (1982).

   Fleeter, R., Kestin, J., Paul, R., Wakeham, W.A.; Physica 108A, 371 (1981).

   Clifford, A.A., Kestin, J., Wakeham, W.A.; Ber. Bunsenges. Phys. Chem. 85, 385 (1981).

9. Diller, D.E.; "Measurements of the viscosity of compressed gaseous and liquid nitrogen + methane mixtures. Intern. J. Thermophys. to be published, 1983.

10. Ely, J.F., Hanley, H.J.M.; Natl. Bur. Stand. Tech. Note 1039 (April 1981).

11. Ely, J.F., Hanley, H.J.M.; Ind. Engr. Chem. Fund. 20, 323 (1981).

12. Ely, J.F., Hanley, H.J.M.; "Prediction of Transport Properties. II. Thermal Conductivity of Pure Fluids and Mixtures", Ind. Engr. Chem. Fund. To be published, 1983.

Acknowledgement

This work was funded by the Office of Standard Reference Data, National Bureau of Standards.

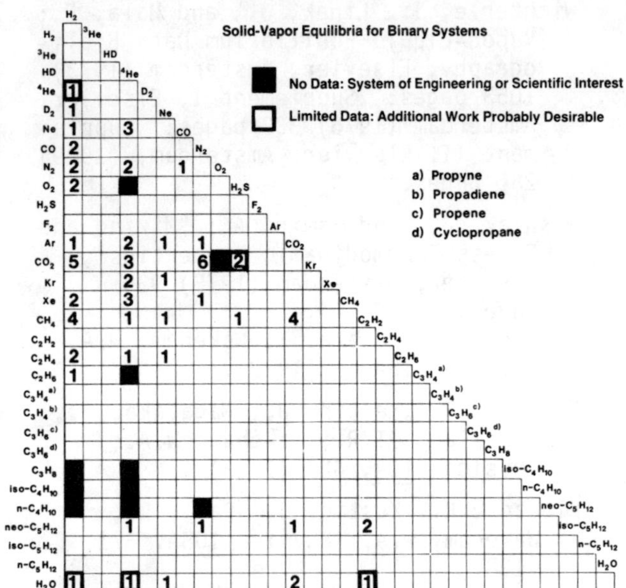

Fig. 1.

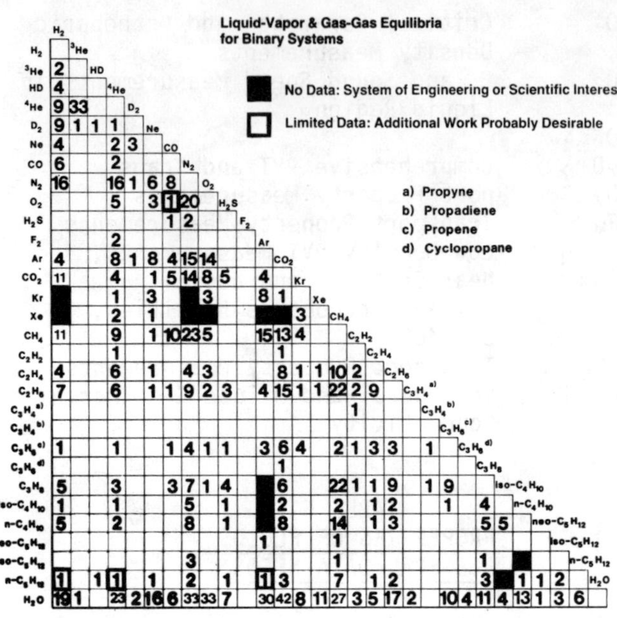

Fig. 3.

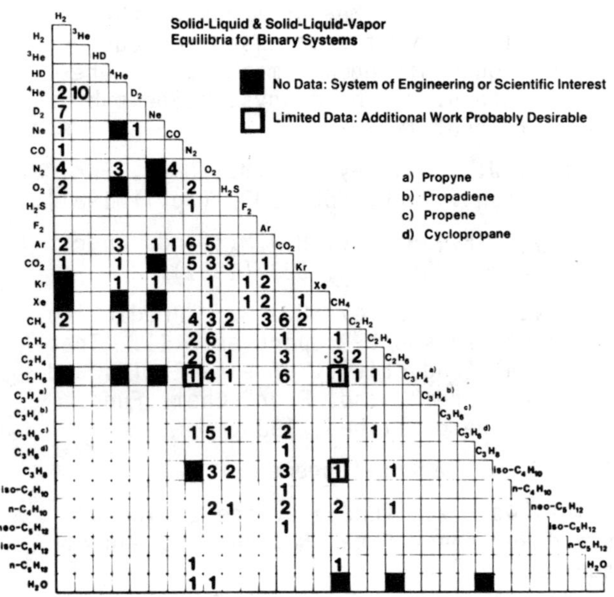

Fig. 2.

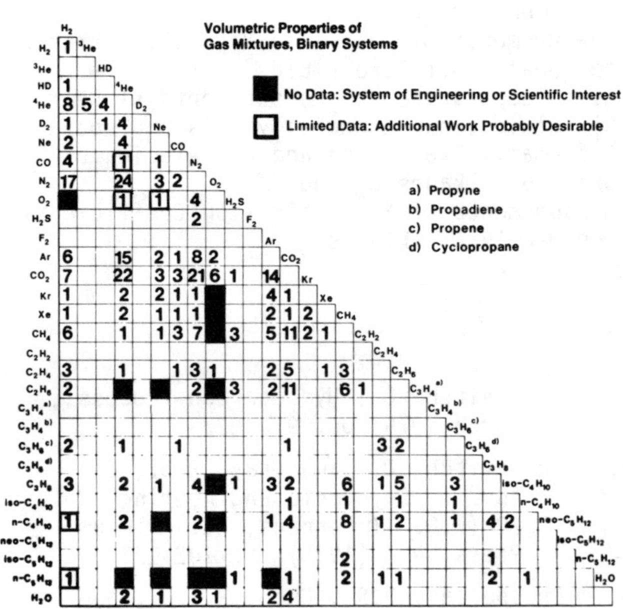

Fig. 4.

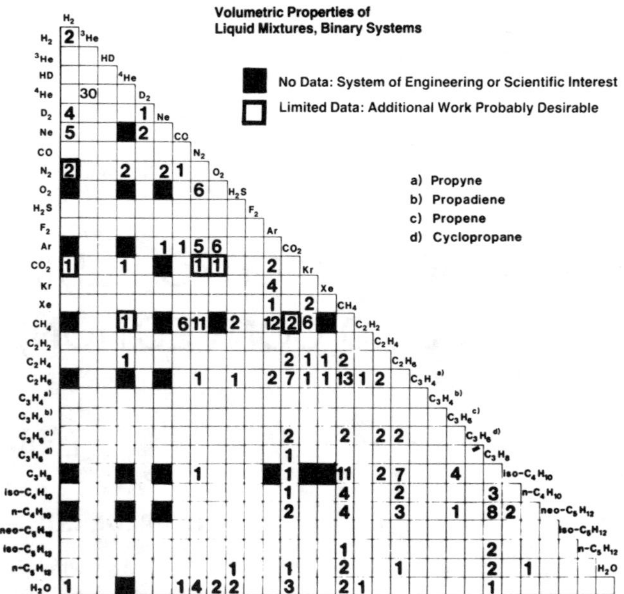
Volumetric Properties of
Liquid Mixtures, Binary Systems
No Data: System of Engineering or Scientific Interest
Limited Data: Additional Work Probably Desirable
a) Propyne
b) Propadiene
c) Propene
d) Cyclopropane

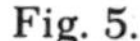
Fig. 5.

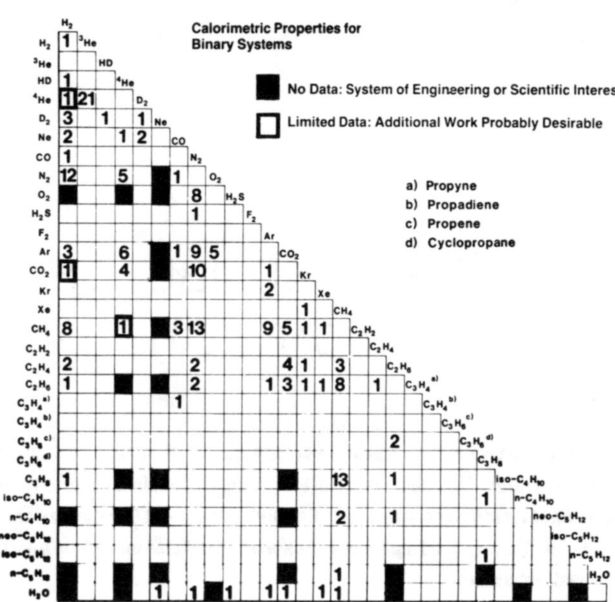
Calorimetric Properties for
Binary Systems
No Data: System of Engineering or Scientific Interest
Limited Data: Additional Work Probably Desirable
a) Propyne
b) Propadiene
c) Propene
d) Cyclopropane

Fig. 6.

IRREVERSIBLE THERMODYNAMICS OF CRYO-FLUID CONVECTION (NEWTONIAN FLUID CONVECTION)

Convection is treated on the basis of the postulate of minimum entropy rejection rates to an entropy sink. This modification of the principle of minimum entropy production rates is illustrated for cellular convection of Rayleigh-Benard and Couette-Taylor flow including kinematics. There is good agreement of data with predicted mean field transport rates and dominant frequencies.

Y. I. KIM
and
T. H. K. FREDERKING
Univ. California
Los Angeles, CA 90024

INTRODUCTION

The concepts of irreversible thermodynamics have been utilized primarily in solid systems, membrane transport and superfluid liquid helium (He II = He^4 below the lambda temperature T). The present study has the purpose of using irreversible thermodynamics to predict convective transport phenomena with emphasis on cryo-fluids. They have small shear viscosities . Therefore convection is established easily. Further, thermal radiation shielding is well developed to exclude contributions to Newtonian fluid transport arising from photon radiation.

After a discussion of the basic postulates the present approach is first illustrated for Rayleigh-Benard convection. Subsequently radial transport in a Couette-Taylor system is treated in the narrow gap limit. The convection kinematics is obtained by an extension of the approach to dominant frequencies and their order of magnitude respectively. Finally conclusions are drawn from the comparison of data with prediction based on the present approach.

BASIC POSTULATES

The following three basic assumptions are introduced: 1. In the fluid *local* thermodynamic equilibrium is established by a sufficient number of collisions involving the entropy carrier system. 2. Thermo-physical transport properties of bulk fluid associated with *very small* departures from thermodynamic equilibrium are valid. 3. The system is assumed to be "*far from equilibrium*" (Prigogine (1)) as far as convection is concerned. This third point refers to the condition of large flow Reynolds numbers (>> 1), and large Rayleigh numbers (>> 1) respectively. In spite of the large Ra number, the first assumption of *local* equilibrium makes it possible to utilize the thermophysical transport properties during equilibration processes.

RAYLEIGH-BENARD CONVECTION

One of the simplest convection geometries is the horizontal Benard layer of fluid (single-phase). Consider a fluid between two horizontal Benard plates (lower plate heated to (T + ΔT), upper plate kept at T by cooling). Fluid between the plates is subjected to a mean heat flux density q supplied at the lower plate and removed at the upper plate. Diffusion takes place up to the critical value Ra_c. At Ra_c convection initiates. A cellular convection pattern appears to be favored energetically. At the order of magnitude 10 Ra_c, the cell pattern may be approximated by a model involving thin boundary layers at the plates. The model assumes that there is an extended domain of inviscid Euler fluid between the boundary layers (with the exception of a center singularity) as seen in the inset of Figure 1. The circulation of the core of Euler fluid is maintained by buoyancy imposed externally due to ΔT in the presence of gravity (and a body force in general). A finite entropy rejection rate is required for steady transport.

There are two major rate contributions: one is the heat supplied by diffusion at the wall; another one is the convective transport rate impeded by viscous forces. According to the principle of minimum entropy production rates (Glansdorff and Prigogine (2)) this quantity should attain a minimum. This principle is modified to the postulate that for steady transport the system equilibration is characterized by a minimum in the entropy rejection rate ($\dot{\sigma}_{tot}$) to the upper plate at the sink temperature T. For very small $\Delta T \ll T$, entropy rejection rates may be replaced by thermal energy rejection rates divided by T. The total contribution is

$$\dot{\sigma}_{tot} = \dot{Q}_c/T + \dot{Q}_k/T \tag{1}$$

(Appendix A). Equation (1) may be rewritten in terms of the boundary layer thickness (mean field value) as

$$\dot{\sigma}_{tot} = B_1\delta^3 + B_2/\delta \tag{2}$$

Minimization of $\dot{\sigma}_{tot}$ with respect to δ leads to an optimum thickness δ^* with a minimum rate

$$\dot{\sigma}_{min} = (4/3)(B_2/\delta^*) \tag{3}$$

Using a reference rate ($sk\Delta T/T$) one may write

$$\dot{\sigma}_{min}/(sk\Delta T/T) = C_\sigma\ Ra^{1/4} \tag{4}$$

The resulting mean field transport rate is expressed in dimensionless form as

$$Nu = C_{Nu}\ Ra^{1/4} \tag{5}$$

Figure 1 shows ($\dot{\sigma}/\dot{\sigma}_{min}$ as a function of the thickness ratio (δ/s). Further, the various (normalized) contributions arising from conduction and convection are included.

Figure 2 presents the influence of the Rayleigh number on the normalized total rejection rate. As Ra is increased, the position of the minimum in $\dot{\sigma}_{tot}$ is decreased as a monotonic function of Ra.

Transport rates Nu(Ra) for liquid He I are in good agreement with Equation (5), and consistent with Equations (3) and (4) (Ahlers (3), Caspi et al. (4)).

OTHER TRANSPORT PHENOMENA

Among other simple flow patterns is cellular transport in a Couette-Taylor system (inner

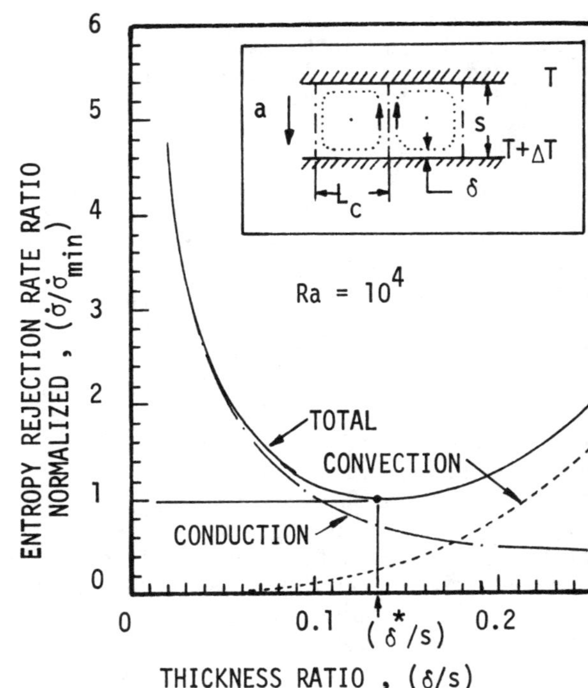

Fig. 1 Entropy rejection rate (normalized) and contributions versus (δ/s); insert: Benard layer (schematic).

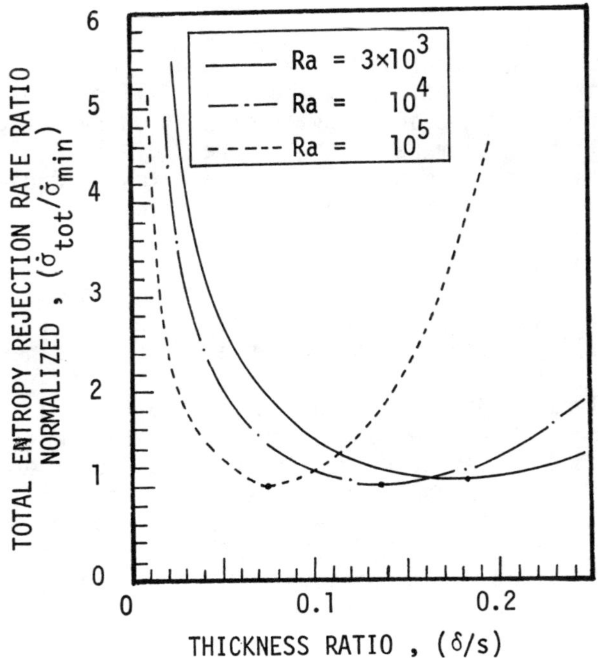

Fig. 2 Total normalized entropy rejection rate versus thickness ratio at various Ra.

cylinder rotating, outer cylinder stationary) (Schlichting (5)). This equilibration case has received attention in conjunction with cryo-fluid transfer tubes for superconducting a.c. generators.

Minimization of the entropy rejection rate for angular momentum transport considers the wall shear contribution and the convection contribution (Kim (6)). Using the present approach an optimum thickness of the wall boundary layer is obtained. Thus, the total entropy rejection rate is minimized again. Often entropy is rejected at the outer wall.

Another case is radial transport of entropy and heat across the fluid layer in a narrow gap of Couette-Taylor flow. For instance heat may be rejected at the outer cylinder in conventional generator cooling systems. Again the sum of the entropy rejection contributions related to $\dot{Q}_k$ and $\dot{Q}_c$ is minimized with respect to the boundary layer thickness δ. The resulting radial Nusselt number is found to be a function of the 4th root of the product of the Taylor number and the Prandtl number. Instead of Nu, one may utilize the dimensionless q-ratio q/q_{ref} with

$$q_{ref} = \eta k^2 \Delta T/(R\Omega^2 \rho^2 s^4) \quad (6)$$

The result obtained for radial heat flow is

$$q/q_{ref} = C_q (\text{Ta Pr})^{5/4} \quad (7)$$

Brunet and Renard (7) have used a Couette-Goertler geometry with the outer cylinder rotating and the inner cylinder stationary using cold N_2 gas or He^4 gas. Figure 3 displays data in the dimensionless coordinates of Equation (7). There appears to be reasonable agreement with the data-compilation for room temperature fluids (6).

A third case without a confining second solid wall is two-phase convection during film boiling. The previous assumption of Boussinesq fluid is relaxed, and the buoyancy term involves large changes in density and temperature. Therefore, application of the present method relies conveniently on the use of thermophysical properties at the mean vapor film temperature. Calculated minimum entropy rejection rate and Nu numbers (Appendix B) are consistent with experiments using horizontal plates cooled by saturated cryo-liquid undergoing film boiling on top. The size, s, of the Benard plate apparatus is replaced by the characteristic size of a vapor bubble B_B. The enthalpy difference $c_p \Delta T$ of Rayleigh-Benard convection is to be replaced by the ΔH from the saturated liquid state to the mean vapor film state.

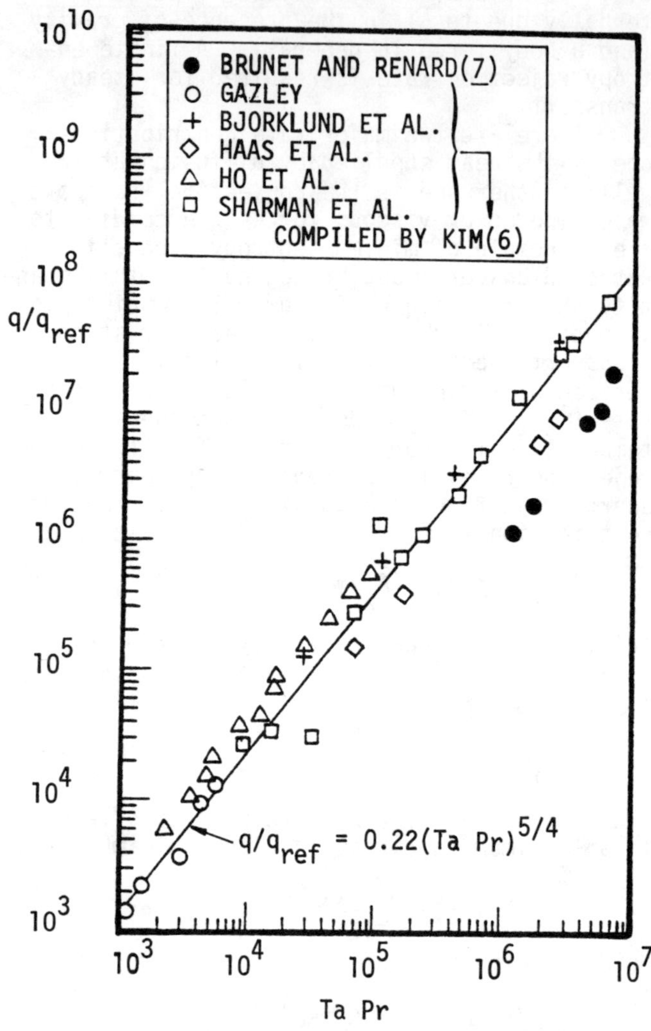

Fig. 3 Dimensionless quantity (q/q_{ref}) versus (Ta Pr) for radial heat transfer in the narrow gap of Couette-Taylor flow.

Other cases of cryo-fluid transport arise in the context of large LNG containers. The open-ended Gorter-Mellink geometry may be useful for the assessment of turbulent transport (Kim et al. (8)). Cold He^4 in a small space may be used conveniently to simulate turbulent flow patterns of large containers at elevated temeratures.

The present results are readily extended to fully established turbulence (Kim (6)). Turbulent axial transport of Couette-Taylor flow has been investigated as modified open-ended Gorter-Mellink transport with one moving wall (insert of Figure 4). The data obtained are described by

$$(k_{eff}/k)(\text{Ta Pr}) = 1.9(\text{Ta Pr})^{4/3} \quad (8)$$

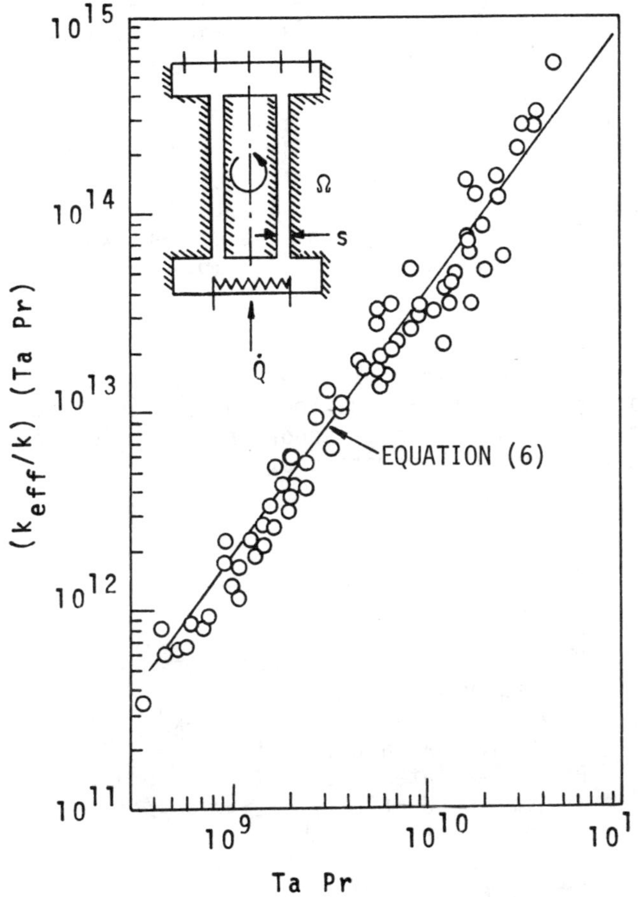

Fig. 4 Data for axial turbulent transport: Open-ended Gorter-Mellink duct with one moving wall (Couette-Taylor case); (Kim (6)) .

The rate results of Figure 4 are considerably lower than data for the stationary Gorter-Mellink case (Kim (6)).

CONVECTION KINEMATICS

The mean flow path ($\sim\pi s$) divided by the residence time ($1/f$) is equal to the characteristic speed. The latter has been deduced from the equation of motion (Appendix A) for conditions of the thin boundary layers of Rayleigh Benard convection. The speed involves the optimum thickness δ^* obtained as Equation (A.5). Therefore, the dominant frequency of the cellular convection regime is readily available as

$$f \sim v_c/s \qquad (9)$$

Using a reference frequency f_{ref} based on the thermal diffusivity, one may rewrite Equation (9) in dimensionless form as

$$f/f_{ref} = C_f Ra^{1/2} \qquad (10)$$

(constant C_f of the order of 0.1).

Figure 5 displays (f_{ref}/f) as a function of Ra. Above Ra_c first a single frequency may be visible, and subsequently two f-values may occur. Figure 5 shows the upper and lower f-function of Gollub-Benson (9). When Ra is increased, however, a transition toward turbulent motion occurs with absence of any dominant length (e.g. plate spacing s). For this condition of locally homogeneous turbulence, Equation (10) has to be replaced by

$$f/f_{ref} = C_{ft} Ra^{2/3} \qquad (11)$$

Because of the turbulence, the frequencies found in this Ra-range may cover a broad band. According to the compilation of Krishnamurti (10) dominant frequencies are indeed characterized by a power law exponent d log f/

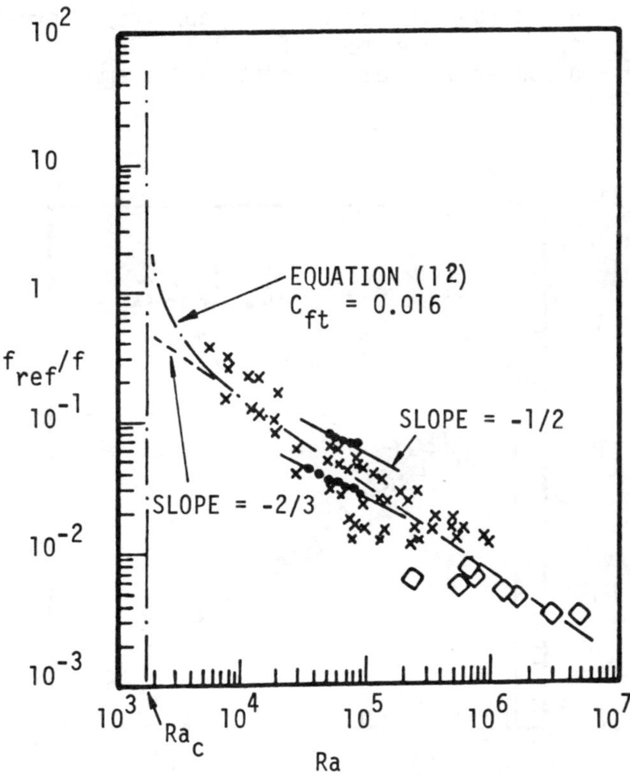

Fig. 5 Dimensionless characteristic time (f_{ref}/f) versus Ra; ● Gollub and Benson (9); × Krishnamurti's (10) compilation; ◇ Kim (6) .

d log Ra = 2/3. He I-data in Figure 5 (Kim (6)) have been taken for pressurized fluid inside an annulus.

In view of the relatively large data scatter, it is convenient to introduce a simple equation (modified Busse equation, Busse (11)),

$$f/f_{ref} = C_{ft}(Ra-Ra_c)^{2/3} \qquad (12)$$

This equation has been included in Figure 5 with C_{ft} = 0.016. It provides a first order approximation to f-data. Equation (12) does not take into account the change in the exponent as one goes from Equation (10) to Equation (11).

Figure 6 presents dimensionless times (= Fourier numbers) obtained during onset of convection studies (Nielsen-Sabersky (12)). A step input in power was applied externally to silicon oil located between Benard plates. At high $\tilde{Ra}$ the exponent (2/3) represents the data well. Equation (12) appears to be a reasonable first order approximation.

CONCLUSIONS

The present approach based on the postulate of minimum entropy rejection rates appears to predict transport phenomena of cellular Rayleigh-Benard convection well. Additional examples pertaining to the narrow gap configuration of Couette-Taylor flow show similarly useful results. These include extensions to locally homogeneous turbulence and to the kinematics of the fluid motion. The method is not restricted to single-phase fluid. The present approach appears to be useful for the two-phase motion of film boiling too.;(Appendix B).

Additional details of the present work have been given elsewhere (Kim (6)).

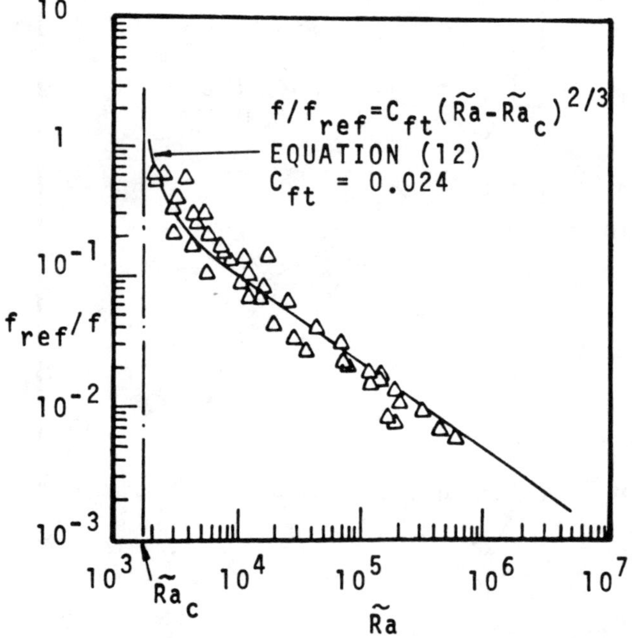

Fig. 6 Dimensionless time needed for the onset of convection for step inputs in power as a function of $\tilde{Ra}$ (Nielsen and Sabersky (12)).

ACKNOWLEDGMENT

This work has been supported in part by the National Science Foundation. We are indebted to Y. Kamioka for his support and contribution in the area of kinematics studies.

NOTATION

B_B	Characteristic bubble size
B_i	Parameter (i = 1,2)
C_i	Constant (i = c,k,f,ft,Nu,q,σ,b,B)
c_p	Specific heat at constant pressure
f	Frequency
f_i	Frequency (i = 1,2,ref); $f_{ref}=k/(\rho c_p s^2)$
g	Gravitational acceleration
ΔH	Effective latent heat
k	Thermal conductivity; $k_{eff} = qL_c/\Delta T$
L_c	Characteristic length
Nu	Nusselt number
P	Pressure
Pr	Prandtl number
q	Heat flux density
q_{ref}	Reference transport rate
Q_i	Heat flux (i = c,k)
R	Radius
Ra	Rayleigh number = $g\beta\Delta T\rho^2 s^3 Pr/\eta^2$
$\tilde{Ra}_c$	Rayleigh number = $g\beta s^4 q\rho^2 c_p/(k^2\eta)$
s	Reference length (Figures 1,4)
t	Time
T	Temperature
T_λ	Lambda transition temperature
Ta	Taylor number = $R\Omega^2\rho^2 s^3/\eta^2$
ΔT	Temperature difference
v	Velocity
v_c	Characteristic speed
β	Isobaric expansion coefficient
η	Shear viscosity
Ω	Angular speed
ρ	Density
θ	Temperature departure from T-profile of conduction
$\dot{\sigma}$	Entropy rejection rate
$\dot{\sigma}_i$	Entropy rejection rate (i = c,k,min,tot)
δ	Boundary layer thickness
δ*	Optimum layer thickness
Δρ	Density difference

Subscript

c Critical, convective
k Conduction, diffusion
min Minimum
ref Reference
tot Total
v Vapor

LITERATURE CITED

1. Prigogine, I., Introduction to Thermodynamics of Irreversible Processes, Wiley, New York (1967).
2. Glansdorff, P., and Prigogine, I., Thermodynamic Theory of Structure, Stability and Fluctuations, Wiley, New York (1971).
3. Ahlers, Guenter, Physical Review Letters, 33, 1185 (1974).
4. Caspi, S., Lee, J. Y., and Frederking, T. H. K., J. Engng. Industry, 99, 723 (1977).
5. Schlichting, Hermann, Boundary Layer Theory, 6th ed., p. 500, McGraw-Hill, New York (1968).
6. Kim, Y. I., "Axial Transport of Heat and Entropy in He I at Low Temperatures through a Stationary Duct and an Annulus with a Rotating Inner Cylinder", Ph.D. thesis, University of California, Los Angeles (1982).
7. Brunet, Y., and Renard, M., Cryogenics, 17, 422 (1977).
8. Kim, Y. I., and Frederking, T. H. K., Chem. Engng. Communications, to be published.
9. Gollub, J. P., and Benson, S. V., Physical Review Letters, 41, 948 (1978).
10. Krishnamurti, Ruby, J. Fluid Mechanics, 42, 309 (1970).
11. Busse, F. H., J. Fluid Mechanics, 52, 97 (1972).
12. Nielson, R. C., and Sabersky, R. H., Int. J. Heat Mass Transfer, 16, 2407 (1973).

APPENDIX A. RAYLEIGH-BENARD CONVECTION

Boussinesq fluid is assumed. Fluid motion between the two horizontal Benard plates is described by the continuity equation (div $\vec{v}$ = 0), and two convection equations:

$$D\vec{v}/Dt = -\nabla P/\rho + g\beta\theta + (\eta/\rho)\nabla^2\vec{v} \qquad (A.1)$$

$$D\theta/Dt = k\nabla^2\theta/(\rho c_p) \qquad (A.2)$$

Boundary conditions: v = 0 at walls and isothermal plates mentioned above.

From the boundary layer-Euler domain model of cellular convection, the heat diffusion rate $\dot{Q}_k$, and the convection term $\dot{Q}_c$ is deduced readily from Equations (A.1) and (A.2). The coefficient of $\dot{Q}_c$ (Equation (2)) is

$$B_1 = C_c s\rho^2 c_p(\Delta T)^2 g\beta/(T\eta) \qquad (A.3)$$

and the diffusion heat-related term is

$$B_2 = C_k s^2 k\Delta T/T \qquad (A.4)$$

The minimization procedure for $\dot{\sigma}_{tot}$ involves $d\dot{\sigma}_{tot}/d\delta = 0$, i.e.

$$\delta^* = (B_2/3B_1)^{1/4} \qquad (A.5)$$

We obtain a ratio $\delta^*/s = Nu^{-1} = (1/0.22)Ra^{-1/4}$ with the coefficient C_{Nu} of Equation (5) evaluated from He I experiments (Ahlers (3)).

The dominant frequency is the reciprocal "residence time" = ratio of the mean flow path to the characteristic convective speed, i.e. $f \sim v_c/(s\pi)$. In dimensionless form based on a reference frequency f_{ref} we obtain

$$f/f_{ref} = C_{ft}Ra^{1/2} \qquad (A.6)$$

(C_{ft} of the order 0.1). The size-independent case of locally homogeneous turbulence is worked out readily. The frequency ratio will be proportional to $Ra^{2/3}$.

APPENDIX B. FILM BOILING

Consider a cryogenic liquid near saturation above a hot horizontal plate with heat flow up. The modified equation of motion is

$$D\vec{v}/Dt = g(\Delta\rho/\rho) + (\eta/\rho)\nabla^2\vec{v} \qquad (B.1)$$

Irreversible thermodynamics is formulated for the thin vapor film which replaces the boundary layer of Appendix A for Benard convection. Minimization of the total entropy rejection rate leads to an optimum thickness

$$\delta^* = C_B[B_B k_v \eta_v \Delta T/(\Delta H \rho_v g\Delta\rho)]^{1/4} \qquad (B.2)$$

where B_B = characteristic bubble size (square root of the surface tension divided by $g\Delta\rho$). The frequency is related to B_B by

$$f(B_B)^{1/2} = C_b(g\Delta\rho/\rho)^{1/2} \qquad (B.3)$$

POOL BOILING HELIUM HEAT TRANSFER FROM MAGNET CONDUCTOR SURFACES

EMMET H. CHRISTENSEN

General Dynamics Convair Division
P.O. Box 80847
San Diego, California 92138

The most common means of cooling and cryostabilizing superconducting magnet windings is by pool-boiling liquid helium. Primarily because of the gravity influences, pool-boiling heat transfer in monolithic, copper-stabilized superconducting windings is a function of the shape, finish, orientation, ventilation paths, helium channel depth, and size of the conductor. The required accurate heat transfer data has generally been obtained from detailed LHe bath heat transfer tests of representative sections of the proposed winding pack, simulating the geometry of the conductors and interconductor insulations of slotted-sheet and pad-on-a-carrier generic types. Availability of data within the magnet community is limited, since the only valid data are those for heat transfer within a simulated winding pack where the test conductors are buried well within the pack, accurately simulating the replenishment hydrodynamics of the helium. To minimize running these complex and expensive tests for future pool-boiling monolithic conductor windings, the existing steady-state boiling heat flux data from those tests have been gathered and correlated in terms of conductor-to-conductor gaps and conductor dimensions. The effects of conductor orientation, proximity of structural walls to the conductor, and influence of surface coatings are discussed in relation to testing and magnet design. These data may be also representative of other LHe boiling processes from surfaces with surrounding constricted ventilation and vapor clearing passages.

INTRODUCTION

Most pool-boiling, helium-cooled magnet, copper stabilized superconductors are cooled by surface contact with helium afforded by passages formed by the insluation in the form of pads-on-a-carrier, slotted sheets, channeled sheets, or barber-pole wraps. Cryostabilization analyses and design requires an accurate boiling heat removal characteristic for the particular winding configuration. To obtain valid heat transfer data for design, it has been necessary to make measurements in bundles of simulated windings with the conductors, the intervening insulation, and helium passages. The conductor's surface form and coating, the orientation of the conductor bundle, and, in some cases, the orientation and form of the insulation passages have been varied in these tests to satisfy conditions at different locations in the magnet. Sufficient tests and data are now available to attempt a unified correlation of these diverse boiling heat transfer data and to generate empirical models for future design. These data are somewhat limited because of the complexity of winding simulated tests, but approximately 10 tests of 16 basic winding configurations have been conducted or monitored by General Dynamics Convair Division. These data are augmented by four other tests encompassing seven more basic configurations. References are provided for the tests and most of the data. Empirical models for the boiling heat transfer are defined in terms of peak nucleate boiling flux or breakaway flux, minimum film boiling flux or recovery flux, and the slope of the film boiling characteristic as illustrated in Figure 1.

LHe BOILING FLUX PROFILE

Important characteristics for boiling heat transfer in magnet windings are depicted by the typical data of Figure 1. Cooling heat flux (W/cm^2) from the wetted surface of the copper conductor to the LHe is plotted versus the temperature rise of the conductor surface above LHe bath temperature. The nucleate boiling flux rises monotonically to the critical peak nucleate boiling or break-away flux which occurs at temperature rises of 0.5 to 0.8K. At higher temperatures there is a temperature zone of boiling transition. The critical minimum film boiling or recovery flux occurs at temperature rises of 2 to 3.5K (3K nominal). The film boiling flux slopes linearly upward at higher temperature rise with the slope defined approximately by a line through a

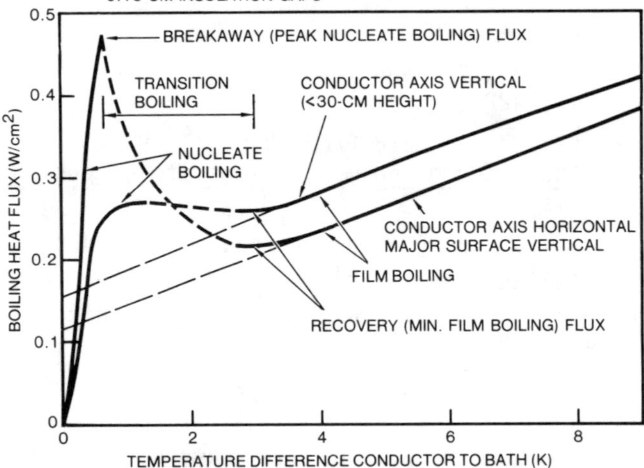

Figure 1. Typical LHe boiling heat flux profile is characterized by critical breakaway and recovery fluxes and film boiling slope.

flux of 0.11 $\pm$ 0.05 W/cm^2 on the ordinate and the minimum film boiling flux point. In the synthesis of the total curve, the flux ordinate intercept and temperature rises of the critical fluxes may generally be nominal values or, can be dispersed within the range in the direction to provide conservatism for the particular usage.

The transition region is characterized by the difference in level of the nucleate boiling maximum and film boiling minimum flux, a hysteresis condition. Hysteresis tends to be nonexistent for long vertical channels where vertically accumulating vapor contributes to breakaway at lower fluxes. The curve in the hysteretic transition zone can be estimated as in Figure 1. For magnet conductor design then, suitable boiling heat flux profiles can be estimated if the two critical fluxes are defined.

CORRELATING EQUATIONS FOR CRITICAL FLUXES

Early tests to determine heat transfer in boiling liquid helium for superconducting magnet cryostabilization were conducted on idealized geometries. Open pool-boiling data, of course, was known to be inapplicable to the small passages in magnets. Several investigators began to look at small, closed-sided channel data with axis vertically oriented. Wilson's data is used as a starting point because the range of his parameters and data are sufficient for magnet conductor applications and to establish the form of the basic correlation equations.[1] Over a wide range of geometry parameters, the breakaway and recovery flux data appear correlatable by an equation of the form:

$$q = C\left(\frac{t}{Z}\right)^{1/2} \qquad (1)$$

where: q is heat flux, W/cm^2
t is the helium gap depth, cm
Z is the vertical height of the channel, cm
C is a constant of proportionality, W/cm^2
Subscripts R or B are used to designate recovery or breakaway.

Wilson's data is for a channel entrance and exit to open bath, with heating on both sides of the channel, exhibiting no hysteresis, and $C = 2.41$ with a maximum error of $\pm 10\%$. The recovery flux constant for magnet windings would be expected to be generally less than this, depending on the tortuosity to flow at the inlet and exit. For vertically oriented conductors, the constant would tend to be higher due to single-side heating of the channel and lateral ventilation passages.

Table 1 summarizes the data and parameters for horizontal conductors. Figure 2 shows recovery heat flux plotted versus the physical geometry parameter $(t/Z)^{1/2}$ for selected boundary flow geometries. The slope of the correlating line is the constant of the equation. The correlating equation parameters and constants for several generic conditions are given in Table 2. As seen in Figure 2, the constants for Eq. 1 fit the data with an error of less than ± 10 percent.

To keep the equation as simple as possible for these numerous complex configurations, it was necessary to change the definition of the parameter Z for discrete ranges of the geometry parameters. For vertical conductors, extreme height between diverters was consistently used for Z and for near-horizontal conductors with major-surface nonhorizontal, major-surface height (conductor wide surface width) was used. For horizontal conductors with major-surface horizontal or for round conductors, half circumference was used for Z. For finned surfaces, only the envelope over re-entrant surfaces was considered for Z.

For both near-horizontal and vertical conductors, recovery flux appeared to correlate best with wetted area weighted mean gap depth used for t, while breakaway flux for near-horizontal conductor correlated best with the minimum gap depth at the perimeter of the conductor used for t. Because of the relatively low height of channels for horizontal conductor, boiling heat flux always exhibited hysteresis, i.e., breakaway flux higher than recovery flux.[2] Figure 3 shows the plot of breakaway heat flux versus the physical geometry factor for selected flow geometries. Similarly to recovery flux, the correlating equation parameters and constants for the several generic conditions are given in Table 2.

Conductors that were tested for angled orientations showed higher critical fluxes for axes tilted 30 to 45 degrees from horizontal or for rectangular conductors rolled about their horizontal axis such that the major surface was inclined 30 to 60 degrees from horizontal.[3] These data are also summarized in Table 1. Again, the width of the wide face of the conductor is used for Z with good correlation.

In recent years, large magnet requirements have dictated long, near-vertical runs of conductor such as in very large diameter solenoids and in Tokamak field coils. Several of the heat transfer tests were directed at measuring heat flux for such winding orientations and at techniques for improving or eliminating the degradation of heat transfer vertically upward in chan-

Table 1. Representative near-horizontal copper conductor individual data and parameters.

Conductor	Dimensions		Turn-To-Turn Insulation	Orientation		Ventilation	Critical Flux Parameters: Recovery			Breakaway			Reference
	Width (cm)	Thickness (cm)		Axis	Large Side		Z (cm)	t (cm)	Flux (W/cm²)	Z (cm)	t (cm)	Flux (W/cm²)	
Large Coil Program (GDC-LCP)	3.12	0.986	Paperdoll	Horiz.	Vert.	Basic*	3.12	0.193	0.252	3.12	0.112	0.424	2
			Pad-on-carrier	30°	Vert.	Basic	3.12	0.193	0.344	3.12	0.112	0.478	2
			Pad-on-carrier	Horiz.	Horiz.	Basic	3.52	0.141	0.136	3.52	0.141	0.238	2
			Tee-pads	Horiz.	Vert.	Open Edge	3.12	0.203	0.42	3.12	0.203	0.61	6
Cask Magnet Prototype System (CMPS)	3.05	0.67	Thin clip	Horiz.	Vert.	Wall	3.05	0.1405	0.17	3.05	0.076	0.27	3
			Thin clip	Horiz.	Horiz.	Wall	3.72	0.1405	0.108	3.72	0.076	0.18	3
			Thick clip	Horiz.	Horiz.	Basic	3.72	0.204	0.139	3.72	0.139	0.25	3
			Thin clip	Horiz.	33-½°	Basic	—	—	—	3.05	0.076	0.414	3
High-Field Test Model Magnet (HFT)	3.45	1.016	Full 30° diverter pads	Horiz.	Vert.	Wall	2.18	0.20	0.275	2.18	0.20	0.43	5
Univ. Tenn. Space Inst. (UTSI)	3.10	0.47	Fishbone channeled strip	Horiz.	Vert.	Open Edge	3.10	0.089	0.25	3.10	0.076	0.40	10
Genl. Elec. Large Coil Segment (GE-LCS)	2.90	1.20	Channeled strands w/ channeled sheet	Horiz.	Horiz.	Wall	3.00	0.30	0.18	3.00	0.30	0.19	7
Elmo Bumpy Torus (EBT)	0.50	0.29	50% barber-pole wrap	Horiz.	Horiz.	Basic	0.79	0.036	0.25	0.79	0.018	0.29	8
Open 7-Strand Cable	0.125 Diam.	—	Disc spacer	Horiz.	—	Basic	0.196	0.016	0.361	0.196	0.016	0.385	9

*Basic means typical configuration at interior of pack.

Table 2. Correlation constants for near horizontal axis copper conductor critical heat fluxes.

Conductor Axis Orientation	Major Surface Orientation	Aspect Ratio	Location	Z*	Recovery Flux: C_R Constant	Wall Degrade (%)	Breakaway Flux: C_B Constant	Wall Degrade (%)
Horizontal	Vertical	>2.0	Mid-pack	Conductor height	1.02		2.22	
		>2.0	Next to wall	Conductor height	0.84	18	1.58	29
Horizontal	Horizontal	>2.0	Mid-pack	Conductor half-perimeter	0.67		1.28	
		>2.0	Next to wall	Conductor half-perimeter	0.55	18	1.05	18
Horizontal	Vertical	>2.0	Insulation edge open to bath	Conductor height	1.57		2.41	
Horizontal	Horizontal	<2.0 to round	Mid-pack	Conductor half-perimeter	1.17		1.89	
30° to 45° inclined from horizontal or	Vertical	>2.0	Mid-pack	Conductor height or major width	1.32		2.58	
Horizontal	Rolled 45° to 60° from horizontal	>2.0	Next to wall	Conductor height or major width	0.88	33	2.00	23
Average wall degrade %						23		23

* Gap depth, t, for recovery flux is wetted area weighted mean gap depth; for breakaway flux is minimum gap depth.

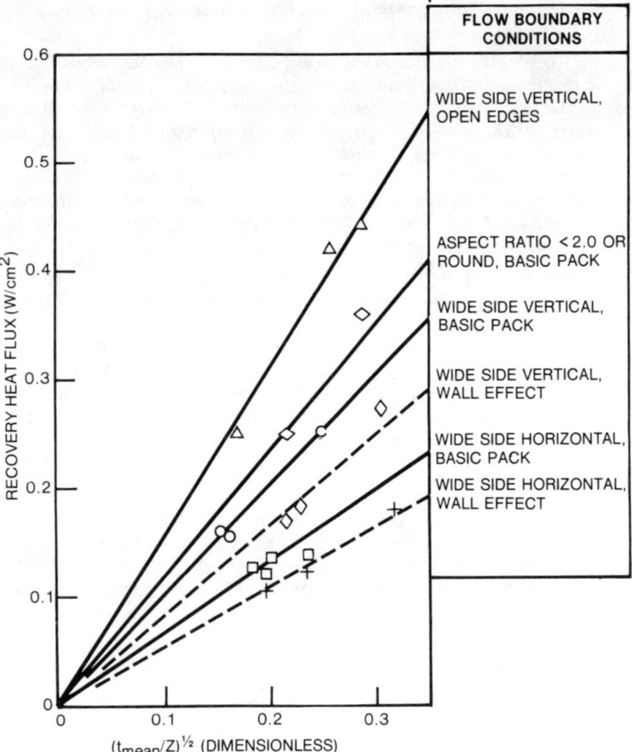

Figure 2. Measured recovery heat flux for horizontal copper conductors correlates with geometry.

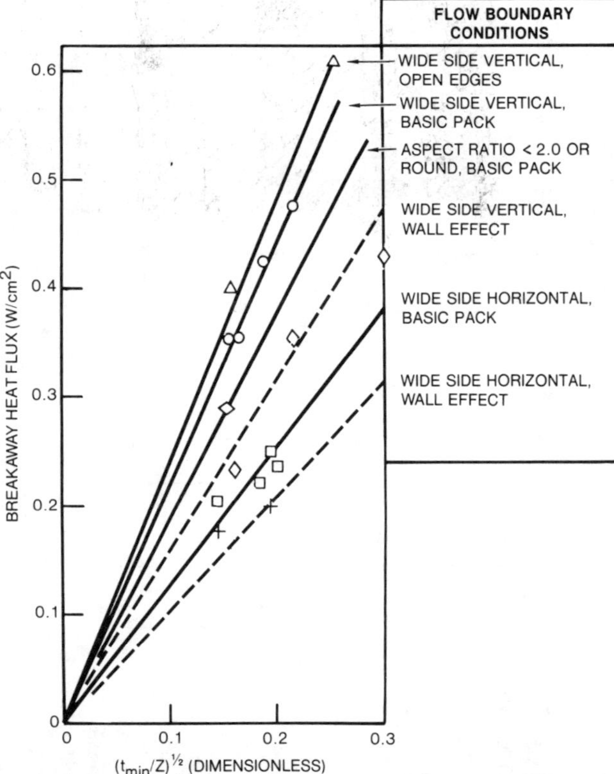

Figure 3. Measured breakaway heat flux for horizontal copper conductors correlates with geometry.

nels parallel to heat dissipating conductor.[2,4,5] The degradation is associated with energy accumulation in the helium rising along the heated conductor. If this is carried too far, the liquid quality decreases to the point of allowing dry-out. Wilson's data addressed this problem, but the heights were limited and the channels closed at the sides. In practical, large magnet windings there is, and must be, some transverse-distributed ventilation.[2] This feature modifies Eq. 1. A modified form that provides good correlation of recovery (and breakaway) heat flux from vertical conductor is found to be:

$$q = C\left(\frac{t}{Z}\right)^{1/2}(1 + 8C_DA^{1/2}) \tag{2}$$

where, in addition to previously defined parameters:

A is the transverse ventilation area to the wide face of a vertical conductor per unit length or height, cm^2/cm

C_D is an effective flow coefficient in the transverse direction, dimensionless.

C_D would be expected to vary with the flow restriction in the transverse direction. For most practical insulations such as a combination of pad-on-a-carrier insulation in one plane and slotted insulation in the other plane, flow restriction might be characterized by two orifices in series and a C_D of about 0.5. For insulations that give a smooth transition from the channels on the side of the conductor to the channels that lead laterally away from the conductor, the C_D could be as high as 1.0.[6] For closed-sided vertical channels, A = 0 and Eq. 2 reduces to the form of Eq. 1. Again, because of the simple form of the correlation equations, the constant is different for different discrete boundary flow geometries.

Figure 4 shows the correlation between recovery heat flux and the modified parameter, $(t/Z)^{1/2}(1 + 8C_DA^{1/2})$ for three vertical conductor channel boundary flow conditions. The constants fit the data for a wide variety of conductors within ±15%.

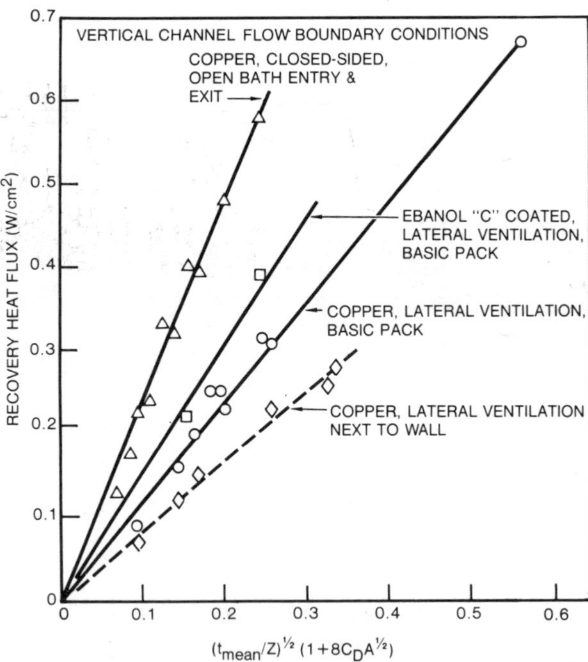

Figure 4. Measured recovery flux for vertical conductors correlates with geometry and lateral ventilation parameters.

For long vertical channels with nondirectional transverse ventilation, boiling heat flux exhibits no hysteresis and breakaway flux is the same as recovery flux.[4] Vertical conductor channels with unidirectional lateral diverters exhibit some hysteretic tendencies because of a component of cross-flow to the conductor similar to helium flow upward across a near-horizontal conductor. Figure 5 shows the general increment of breakaway flux above recovery flux for vertical channels as a function of the unidirectional diverter interval normalized as the ratio of interval height to conductor wide surface width, Z/W. This permits determination of breakaway flux from the recovery flux for these unidirectional diverted configurations.

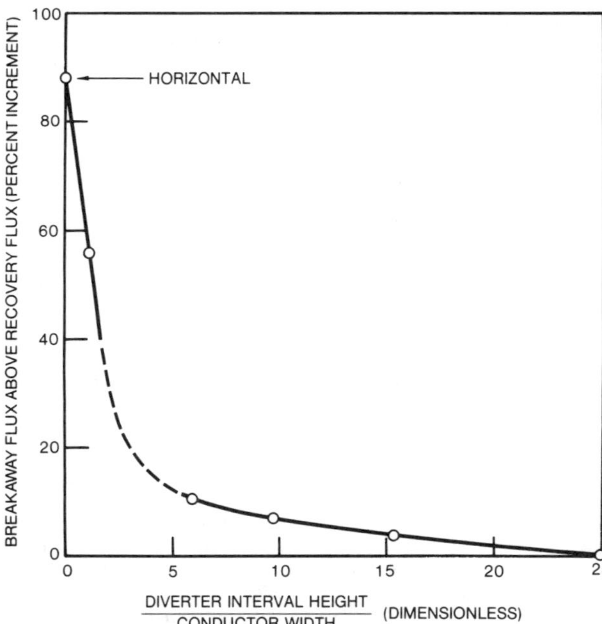

Figure 5. Breakaway flux for vertical conductors approaches recovery flux as uni-directional lateral diversion height increases.

GENERAL VARIATION OF FLUXES WITH PHYSICAL AND BOUNDARY FLOW GEOMETRY

A somewhat less rigorous presentation of correlation is visually meaningful if the 3.12 cm ± 10% variation in the height or major surface width of the first five selected large conductors of Table 1 is ignored and recovery flux plotted versus only wetted area weighted mean gap depth. These conductors have width-to-thickness aspect ratios varying from 2.5 to 6.5. This correlation is shown in Figure 6 and visually exhibits the parabolic characteristic corresponding to $t^{1/2}$ in the correlation equations. Similar, but higher, flux curves (not shown) result when breakaway fluxes are plotted versus minimum gap depth at the conductor envelope.

BOILING CRITICAL FLUXES FOR RANGE OF NONVERTICAL CONDUCTORS

The correlations shown in Table 2 for near-horizontal axis copper conductor demostrate two important cryostability design considerations in pool-boiling superconducting magnets. One is that heat flux is degraded on conductors against walls of the magnet when the wall insulation is channeled to the same depths as the basic interconductor insulations.[4] This degradation is an average of 23% for a broad range of orientations and insulation forms and is due to the partial obstruction by the adjacent wall of the normal three-dimensional helium flow away from and to the heated conductor. Testing demonstrated that the wall proximity conductor heat transfer could be returned to the basic winding recovery heat fluxes by deepening the wall insulation channels.[4] As a matter of fact, the wall insulation channel depth necessary is that which yields a wetted area weighted mean gap depth that, in turn, yields an Eq. 1 (with degraded recovery flux constant) flux equivalent to that deep within the pack. For typical rectangular conductor with 3:1 aspect ratio, flat against the magnet wall, Eq. 1 yields required wall channel depth 4.5 times that between conductors, which correlates well with the test results of References 4 and 8.

The other major result suggested by the correlations of Table 2, and directly quantified in Figure 6, is that wide, flat, rectangular conductor, with the wide surface horizontal, suffers severely degraded heat flux relative to conductor with major surface vertical.[3,7] This horizontal orientation of major surfaces penalty largely disappears for smaller conductors with aspect ratios less than 2.0 and for round conductors.[8,9] The magnet design and fabrication implications are important in that vertical plane, large solenoids with rectangular conductor wound the "easy" way present the degrading orientation at the top and bottom of the magnet. To avoid this, and provide the major surface with a vertical orientation, requires winding the rectangular conductor the "hard" way.

By solving the correlation equations with the parameters of Table 1 or, more directly, by looking at Figure 6, conductor and surfaces inclined from horizontal or vertical result in better critical heat fluxes. Rectangular conductor longitudinal axis inclined 30 to 45 degress from horizontal, or the conductor rolled 30 to 45 degrees around a horizontally oriented longitudinal axis, yields the highest critical fluxes.[2,3] Recovery flux is 30 percent higher than for conductor with horizontal axis and 10 to 18 percent higher than for conductor with vertical axis.

A significant result is that the correlation for small round[9] and low-aspect (<2.0) ratio[8] rectangular conductors, even with major-surface horizontal, shows very high heat fluxes and, when the conductors are small, the helium gaps can also be small. The correlation constant is considerably higher than for the larger conductor with similar gap-to-height ratio (t/Z). This effect for boiling is quite similar to the thin boundary layer effects on small horizontal wires in natural convective environments. Gaps should be no less than 0.030 cm for even small conductors to avoid vapor lock effects, which are manifested as degraded heat transfer.

For horizontal-axis conductor with wide surface horizontal, Figure 2 and Table 2 simply categorize the averaged equation constants by whether width, W, to height, H, aspect ratio is greater or less than 2.0. To be most useful for magnet conductor design and trade studies, a continuous functional variation with aspect ratio is desired. Limited data for horizontal

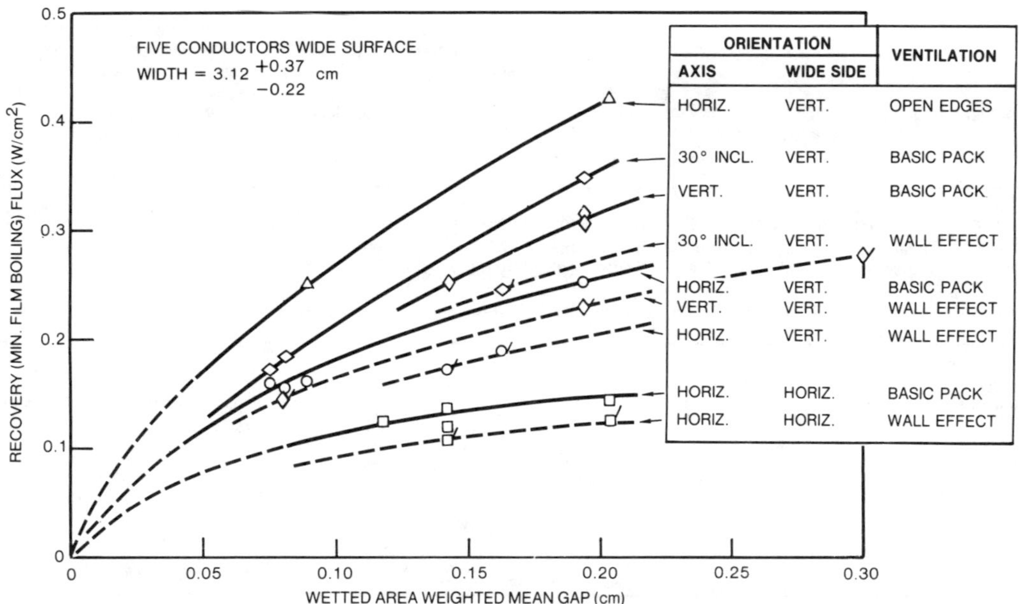

Figure 6. Recovery flux of large conductors varies with helium gap between conductors and orientation/ventilation.

longitudinal axis and horizontal wide surface yield detailed versions of Eq. 1 including the functional-relationship-to-aspect ratio for recovery heat flux. For mid- or basic pack:

$$q_{rcy} = 1.25 \left(\frac{H}{W}\right)^{1/4} \left(\frac{t_{avg}}{W + H}\right)^{1/2} \tag{3}$$

For next-to-wall:

$$q_{rcy} = 0.80 \left(\frac{H}{W}\right)^{1/6} \left(\frac{t_{avg}}{W + H}\right)^{1/2} \tag{4}$$

where, in addition to previously defined parameters:

q_{rcy} is recovery or minimum film boiling flux, W/cm^2.

H is height of horizontal rectangular conductor, cm.
W is width of horizontal rectangular conductor, cm.
Eq. 3 and 4 are valid for horizontal-axis conductor with wide-surface horizontal over an aspect ratio range of $1.0 \leqslant W/H \leqslant 10.0$.

BOILING CRITICAL FLUXES FOR RANGE OF VERTICAL CONDUCTORS

Table 3 shows the nature of the correlation parameters and the correlation constants for vertical channels and conductors. The first entry is for closed-sided vertical channels with open-bath entry and exit, and is a limiting target for practical magnet conductor channels.[1] No hysteresis is exhibited by such channels; i.e., breakaway and recovery flux are identical.

Heat flux degrades upward in vertical channels heated from the side due to the accumulation of energy in the rising helium and results in decreasing average fluxes for the channel with increasing height. This is accounted for by the square root of height, $Z^{1/2}$, in the denominator of the flux correlation equation. The vertical degradation is mitigated by provision for lateral or transverse ventilation in practical magnet windings, but there are still limits to the permissible vertical height of helium rise.[4] The energy-laden helium must be positively diverted laterally, and helium induced to flow diagonally, across the surfaces of the vertically oriented conductor. The data correlated are for vertical conductors and diverter vertical intervals up to 107 cm. Practical positive diversion intervals should generally be less than 20 cm, even for large conductors, to avoid excessive vertical degradation, and should be unidirectional for orderly flow across the two major faces of the conductor. The correlation of the various recovery flux data is shown in Figure 4, including coated conductor; the correlation constants are given in Table 3; the quantitative display of recovery flux for selected conductors are shown in Figure 6.

Rapid lateral diversion in short distances on vertical conductors would be expected to exhibit hysteresis similar to the gravity-driven cross-flow on horizontal conductors. Long, vertical rises before diversion would be expected to approach the nonhysteretic characteristics of "long" vertical, small-gap, closed-sided channels. Breakaway flux for vertical conductor channels is plotted in Figure 5 as percent increase of breakaway flux over recovery flux versus the ratio of height between lateral diversion to the major-surface width of the conductor. Such practical vertical channels with lateral ventilation are virtually nonhysteretic for a ratio of 10 corresponding to 30 cm height on a 3-cm-wide conductor.

HEAT TRANSFER EFFECTS OF SURFACE COATINGS

Insulating coatings of appropriate combinations of thickness and thermal conductivity applied to metals can enhance pool-boiling heat transfer.[3] This phenomenon was applied to the tempering of sword blades for centuries and has been applied to cooldown of missile cryogenic ducts for decades. Table 4 summarizes the critical heat fluxes for lead/tin solder-coated copper and for copper oxide (typified by the trade name Ebanol "C," Ethone Corp.), coated copper. These are compared to bare copper fluxes for corresponding identical geometric configurations.

The performance of solder coating is of great interest to superconductors because solder coating is often unavoidable during fabrication and joining of superconductor matrix to the copper stabilizer. Table 4 shows that recovery fluxes for solder are not detectably different from bare copper within a sensitivity level of ± 10 percent. At the higher breakaway fluxes, increased temperature gradient across the thermal impedance of the solder yields significantly higher breakaway fluxes by 25 to 45 percent depending on orientation. For steady-state cryostability of superconducting magnets, recovery flux is of more significance than breakaway flux and, therefore, solder coating is of little benefit but, more importantly, is not detrimental.

The thin insulating surface coating of copper oxide (Ebanol "C") drastically enhances heat transfer. Recovery heat flux is improved generally 22 to 32 percent while, in the particularly detrimental major surface horizontal

Table 3. Correlation constants for vertical axis conductor critical heat fluxes.

Vertical Channel Transverse Ventilation Configuration	Conductor Surface	Location	Z*	A	Recovery** Flux Constant (C_R)
Closed sides	Bare copper	Entrance and exit open to bath	Height of closed-sided channel	O	2.41
Transverse ventilated sides	Bare copper	Mid-pack	Vertical length of conductor or height between lateral diverters	All transverse area if omni-directional. Half transverse area if uni-directional transverse diversion.	1.20
Transverse ventilated sides	Bare copper	Next-to-wall	Vertical length of conductor or height between lateral diverters	All transverse area if omni-directional. Half transverse area if unidirectional transverse diversion.	0.84
Transverse ventilated sides	Ebanol "C" coated copper	Mid-pack	Vertical length of conductor (30.5 cm)	All transverse area if omni-directional. Half transverse area if unidirectional transverse diversion.	1.54

* Gap depth, t, for recovery flux is wetted area weighted mean gap depth; for breakaway flux is minimum gap depth.

** Breakaway flux relative to recovery flux for unidirectional lateral diversion is a function of the ratios of channel height between lateral diverters to conductor or channel wide face width as given by Figure 5.

Table 4. Correlation constants for near-horizontal axis conductor with coatings.

Conductor Axis Orientation	Major Surface Orientation	Coating	Location	Z*	Recovery Flux: C_R Constant	Recovery Flux: Chg. From Bare Copper (%)	Breakaway Flux: C_B Constant	Breakaway Flux: Chg. From Bare Copper (%)
Horizontal	Vertical	Solder	Next-to-wall	Conductor height	0.80	−5	2.27	+44
Horizontal	34° roll from horizontal	Solder	Next-to-wall	Conductor height	0.80	−9	2.50	+25
Horizontal	Horizontal	Solder	Next-to-wall	Conductor half-perimeter	0.60	+9	1.53	+46
Horizontal	Vertical	Ebanol C	Mid-pack	Conductor height	1.35	+32	2.65	+19
30° from horizontal inclined	Vertical	Ebanol C	Mid-pack	Conductor height	1.69	+28	2.98	+16
Horizontal	Horizontal	Ebanol C	No wall effect	Conductor half-perimeter	0.97	+45	2.42	+89
Horizontal	Horizontal	Ebanol C	Next-to-wall	Conductor half-perimeter	0.83	+51	1.53	+46
Horizontal	Vertical	Ebanol C	Next-to-wall	Conductor height	1.03	+23	2.65	+67
Horizontal	57° roll from horizontal	Ebanol C	Next-to-wall	Conductor height	1.07	+22	2.98	+49

*Gap depth, t, for recovery flux, is wetted area weighted mean gap depth; for breakaway flux is minimum gap depth.

orientation, the improvement is an extremely beneficial 45 to 51 percent. Breakaway flux is enhanced 16 to 19 percent for generally unobstructed configurations, as much as 89 percent for the detrimental horizontal major surface configuration, and an average of 54 percent for wall-proximity-degraded configurations.

Therefore, based on recovery flux, such an insulating coating can easily increase steady-state cryostability by 25 percent for already thermally beneficial orientations, and by 50 percent for thermally detrimental orientations necessitated by inherent or intrinsic winding requirements.

HEAT TRANSFER EFFECTS OF SURFACE FORM

During the referenced testing, in addition to flat-surfaced rectangular conductor, heat transfer on multiple V-groove fins for several conductor axis orientations[2,4,6] and on multiple rectangular fins for vertical conductor were measured.[5] The V-groove fins were parallel to the conductor axis and had a 30-degree included angle, with opening between of 0.19 cm depth, 0.032 cm root gap width, and 0.118 cm opening width. The rectangular fins were also parallel to the conductor axis and were 0.20 cm wide, with opening between of 0.15 cm depth and 0.20 cm width.

Since the rectangular grooves were not run for near-horizontal configurations, comparison can be made between all three surface forms only for vertical and diverted vertical configurations. In general, Eq. 1 and 2 can predict the recovery flux and breakaway flux of the fins alone. If the recovery and breakaway constants of proportionality for the total conductor dimensions are used, then recovery flux t is characterized by the mean gap between adjacent fin surfaces. For the V-groove fins and rectangular fins, recovery t is 0.075 cm and 0.20 cm respectively. Recovery heat flux is 55 percent and 85 percent of flat conductor flux for the V-groove fins and rectangular fins, respectively.

When the surface area enhancement factors of 2.8 for the V-groove fins and 2.00 for the rectangular fins are included, the conductor heat transfer rate is higher than for a flat conductor, by 54 percent and 70 percent respectively. In this respect, rectangular fins are best for vertical conductor. For the vertical fins, little hysteresis is present for any diverted configuration, and breakaway flux is virtually identical to recovery flux.

For horizontal conductor, recovery flux of the V-groove fins is controlled by the same value of $t = 0.075$ cm in Eq. 1. The side of the horizontal grooves is slanted at 15 degrees to the horizontal, which promotes flow of vapor to the interconductor channels. While the quantitative effect was not measured, the rectangular fins in the horizontal orientation would present a downward facing surface and not be as effective in vapor dispersal as the V-grooves. With the basic conductor proportionality constant in Eq. 1, the breakaway flux correlation for the near-horizontal conductor with V-groove fins dominating the surface is controlled by the minimum gap t in the groove, the 0.032-cm root gap. This yields a relatively low breakaway flux for the fins.

CONCLUSIONS

A unified theory and equation correlates LHe boiling critical heat fluxes in a variety of winding and conductor configurations with conductor and passage geometry parameters, through constants of proportionality which vary with the freedom of natural ventilation or boundary flow geometries. This correlation of the data of a wide range of simulated winding pack configurations and tests permits synthesis of design pool-boiling steady-state heat flux characteristics for future windings. Conductors, insulations, and helium passages may be configured through the use of the equations and defined constants to achieve required levels of heat flux and cryostability margins.

Degradation tendencies associated with tortuous ventilation paths, particularly adverse orientations of conductors, and wall effects are defined and quantified. Of even more importance, design techniques are quantified and presented to mitigate, eliminate, or compensate these degradations.

The degree of benefit of two representative coatings in enhancing heat transfer is quantified. The metallic coating of lead/tin solder over copper, which is unavoidable on some conductors, yields at least as good heat flux as bare copper. Thin, thermally insulating coatings represented by copper oxide significantly enhance the pool boiling heat transfer. Insulating coatings have provided enhanced heat transfer in magnets where improved heat transfer and cryostability were required.

Modified surface form of the conductor such as fins can enhance heat transfer and cryostability. Near-optimized configuration V-groove and rectangular fins are similar in overall benefit. The lower, nonhorizontal orientation heat fluxes of V-groove fins relative to rectangular fins is offset by the increased surface area achievable with V-groove fins.

REFERENCES

1. Wilson, M. N., "Heat Transfer to Boiling Liquid Helium in Narrow Vertical Channels," Paper II-2, IIR Commission 1 Meeting, June 1966.
2. Christensen, E. H., "Pool Boiling Helium Heat Transfer in Typical Conductor Packs," Proc. of the 8th Symposium on Engineering Problems of Fusion Research, Vol IV, 1979, pp 1769-1773.
3. Taylor, W. D., J.E.C. Williams, and M. Sinclair, "Pool Boiling LHe Heat Transfer in an MHD Conductor Pack," Paper JA-5, 1981 Cryogenic Engineering Conference, San Diego, CA, 10-14 August 1981.
4. Christensen, E. H. and S. D. Peck, "Pool-Boiling Liquid Helium Heat Transfer in Vertical Conductor Packs," Paper JB-2, 1981 Cryogenic Engineering Conference, San Diego, CA, 10-14 August 1981.
5. Peck, S. D. and E. H. Christensen, "Pool-Boiling Helium Heat Transfer from Two Large Monolithic Copper Stabilizers," Proc. of the 9th Symposium on Engineering Problems of Fusion Research, Vol. I, 1981, pp 277-280.
6. Krause, R. P., E. H. Christensen, R. D. Bradshaw, and R. E. Tatro, "Tests in Helium to Verify Cryostability and Replenishment," IEEE Transactions on Magnetics, Vol. MAG-15, No. 1, Jan 1979, pp 748-751.
7. Michaelson, P. F., R. Quay, R. F. Koenig, P. L. Walstrom, and J. S. Goddard, "Heat Transfer and Helium Replenishment in Cabled Conductor Cooling Channels," Advances in Cryogenic Engineering, Vol. 25, 1980, pp 398-405.
8. Kim, I.-K. and S. D. Peck, "Pool Boiling Helium Heat Transfer from Small Monolithic Copper Stabilizers with High Packing Factor," Paper CG1-2, ICEC9-ICMC, Kobe, Japan, 11-14 May 1982.
9. Khalil, A., "Boiling Heat Transfer on Bundled Conductors in Normal Helium," Paper MA-3, 1981 Cryogenic Engineering Conference, San Diego, CA, 10-14 August 1981.
10. Chen, C.-J., S.-T. Wang, and J. W. Dawson, "Vapor Locking and Heat Transfer Under Transient and Steady-State Conditions," Advances in Cryogenic Engineering, Vol. 25, 1980, pp 412-419.

TRANSIENT HEAT TRANSFER PROCESSES

V. D. ARP
D. E. DANEY
P. J. GIARRATANO
and
W. G. STEWARD

Chemical Engineering Science Division
Center for Chemical Engineering
National Bureau of Standards
Boulder, Colorado 80303

Heat transfer processes which change rapidly with time occur in a wide variety of applications. Examples occur in nuclear reactor safety analysis, internal combustion engine thermal losses, helium cooled superconducting cables, exhaust gas heat recovery, quenching of liquid metals, and other technologies.

Heat transfer design and analysis of such processes customarily starts with a quasi-static approximation, which is defined in the following sense. Let $h_1(a, b, c...)$ be a heat transfer correlation, with a, b, c... representing relevant parameters such as wall temperature, bulk fluid temperature, convective velocity, heat flux, etc. Such correlations are almost always developed from experimental data wherein the parameters a, b, c... are time-independent as each data point is recorded. If now one applies this correlating equation to a situation where, for example, the parameter "a" is rapidly time varying, the quasi-static approximation is that the form of the correlating equation is unchanged, i.e., the correlation is $h_1(a(t), b, c...)$.

In this paper we survey several processes where the quasi-static approximation is invalid. Examples include transient boiling phenomena and transient convective motions in a compressible fluid, both perpendicular and parallel to a heated surface. Experimental measurements have been supported by numerical modeling studies.

Conclusions and Significance

Measurements of time-varying temperatures and heat fluxes, taken with as short as 0.5 microsecond resolution, enable the development of dynamic heat transfer processes to be traced experimentally. The measurements are combined with numerical modeling of the process to provide useful physical models.

Studies have been done on boiling fluids, where it is found that peak nucleate boiling heat transfer rate can be exceeded by factors of 2 or 3 for short (millisecond) times without the occurrence of a transition to film boiling. The delay in the transition to film boiling is consistent with the existence of a thin liquid sublayer at the surface during nucleate boiling.

Two cases of transient heat transfer to a non-boiling fluid have been studied. For steady state turbulent flow in a tube, the effective thermal diffusivity is a function of distance from the tube wall. An analytical solution to the transient diffusion problem has been obtained, as a function of Reynolds number for the fluid flow. The solution agrees well with some experimental measurements in helium and disagrees with some for water, apparently because of finite wall heat capacity effects. For a compressible fluid, fluid motion due to thermal expansion in the boundary layer causes deviations from the usual incompressible fluid thermal diffusion problem. The deviations are generally small for ideal gases, but may be substantial for a near-critical fluid, at least up to the time that buoyant convection begins to contribute.

Numerical modeling has been utilized in a number of these studies, starting with differential equations for non-ideal fluid flow. With a finite compressibility (sound velocity) the equations are hyperbolic, and mathematical stability of the integration routines proves to be a severe problem. When the conservation of mass, momentum, and energy is written using density, temperature, and velocity coordinates, the integration is most often terminated by mathematical instabilities. When the same problem is described using pressure, enthalpy, and velocity coordinates, stable integration is much easier to achieve. We suggest that these latter coordinates are particularly useful outside of the ideal gas or incompressible fluid regimes.

Experimental Techniques

Experimental techniques used in this study have been documented elsewhere, and only a brief summary is presented here.

For studies at liquid helium temperature, a very thin ($\sim 0.5 \times 10^{-6}$m) carbon film deposited on quartz substrate has served as a simultaneous heater and thermometer [1], [2]. Its heat capacity is negligible, and thermal diffusivity from the carbon into the quartz can be neglected below 10 K. Thus, the experimental corrections and calibration are simple, but the experimental interpretation for classical heat transfer may be confused by the "Kapitza" thermal resistance at helium temperatures, and perhaps by the extremely low thermal conductivity of the carbon in the plane of the film, which may affect nucleation processes.

For studies at temperatures of 77 K to nominally 400 K, a similar heater-thermometer consisting of thin ($\sim 10^{-7}$m) platinum film on a quartz substrate has been used for some of the measurements [3]. The heat capacity of the platinum is still very small, but a very substantial correction for time-dependent heat flow into the quartz is required when the fluid density is low. In order to make this correction, the temperature-dependent thermal diffusivity of the quartz is first determined by a transient technique with the sample in a vacuum. Then for each heat transfer experiment this time-dependent correction is computed numerically by integration of the thermal diffusivity equation

$$\frac{\partial T}{\partial t} = \alpha(T) \frac{\partial^2 T}{\partial x^2}$$

using experimentally determined T(t) on both the front and back sides of the quartz as boundary conditions.

Also for studies above 77 K, a 2.5×10^{-5}m platinum wire has served as a heater thermometer for some measurements. The heat capacity of the wire is easily included in the analyses, but the high curvature of the heat transfer surface limits its application.

An optical system for direct measurement of temperature profiles in the heat transfer fluid is under development. This will use interferometric and/or Schlierin techniques, from which temperature gradients can be determined. Preliminary data have been taken, but no results from this technique are available at this time.

Boiling Phenomena

Failure of the quasi-static approximation for transient boiling correlations is related to the time required for bubble nucleation and growth.

For some microseconds after start of a heat pulse to the thermometer, the heat transfer rate is determined primarily by thermal diffusion into a very thin boundary layer. The exception is in the case of liquid helium, where a dominant temperature discontinuity exists between the solid (heater) surface and the fluid boundary layer (the Kapitza resistance); this temperature discontinuity is in series with the temperature gradients associated with thermal diffusion in the boundary layer. However, nucleate boiling heat transfer may be reached in as little as 10^{-5} to 10^{-4} seconds as the heat flux is increased. As the wall temperature and heat flux rise above the "peak nucleate boiling" limit as determined from steady state measurements, the experimental data follow an extension of the nucleate boiling curve for several milliseconds, or until the "peak nucleate boiling" heat flux is exceeded by a factor of two to three. Examples are seen in figures 1 and 2. Finally a rapid transition to film boiling occurs. In our experiments with low thermal inertia thermometers, we have found that the time required to reach film boiling is consistent with a model of a thin liquid sublayer at the heater surface under nucleate boiling conditions, as suggested by [2]. The liquid film thickness is estimated to be ~4 micrometers. For most situations, however, this transition time would be strongly influenced by the thermal inertia of the solid surface.

Local Heat Transfer in a Non-boiling Fluid

We consider here the situation where the fluid velocity parallel to the heat transfer surface is essentially unperturbed by the heat transfer process. This allows consideration of a simple one-dimensional analyses of fluid temperature and movement perpendicular to the heat transfer surface.

After the start of a heat pulse, or thermal perturbation, the heat transfer rate is governed by thermal diffusion processes within a time-varying boundary layer. For a short time, and/or with low heat fluxes, the classical incompressible fluid analyses apply; such problems are discussed in Carslaw and Jaeger [4], for example, and are easily solved with modern numerical techniques. If the fluid is static, or in laminar flow, and is indeed rather incompressible, there remains little to perturb this solution. If the fluid is in turbulent flow, then the effective thermal diffusivity is dependent upon the turbulence profile, and thus varies with distance from the heat transfer surface.

Following the procedure of Reynolds, Von Karman, and Prandtl, an eddy diffusivity, (ε_h) is obtained which transports heat through the boundary layer in addition to that transported by the molecular diffusivity α. This eddy diffusivity ε_h is a consequence of the turbulence fluctuations rather than a property of the fluid. The eddy diffusivity is given [5] as

$$\varepsilon_h = \xi y,$$

where

$$\xi = 0.4U_m \sqrt{\frac{Nu}{Re}} \text{ (steady state)}$$

y is the distance from the surface, U_m is the mean velocity, Nu is the Nusselt number, and Re is the Reynolds number. The Fourier equation modified by the combined diffusivity provides a description of the transient temperature field emanating from a flat plate.

$$\frac{\partial}{\partial y}(\alpha + \xi y)\frac{\partial T}{\partial y} = \frac{\partial T}{\partial t}$$

With the boundary condition for a step heat input into the fluid, and a steady bulk temperature T_b at depth, δ, a closed solution for the wall temperature T_w can be obtained,

$$\frac{T_w}{T_{w,ss}} = 1 + \sum_{n=1}^{\infty} b_n e^{-\left(\frac{\gamma_n}{c}\right)^2 t^*},$$

where $T_{w,ss}$ is the steady state wall temperature, where γ_n are the roots of

$$J_1\left(\frac{\gamma_n}{c}\right) Y_0(\gamma_n) - Y_1\left(\frac{\gamma_n}{c}\right) J_0(\gamma_n) = 0;$$

$$b_n = \frac{\pi c}{\ell n\, c}\,\frac{J_0^{\,2}(\gamma_n)\left[J_0\left(\frac{\gamma_n}{c}\right)Y_1\left(\frac{\gamma_n}{c}\right) - Y_0\left(\frac{\gamma_n}{c}\right)J_1\left(\frac{\gamma_n}{c}\right)\right]}{\gamma_n\left[J_1^{\,2}\left(\frac{\gamma_n}{c}\right) - J_0^{\,2}(\gamma_n)\right]};$$

J_0 and J_1 are Bessel functions of the first kind, order 0 and 1; Y_0 and Y_1 are Bessel functions of the second kind, order 0 and 1.

$$t^* = \frac{\xi^2 t}{4\alpha} \quad \text{(dimensionless time)}$$

and

$$c = \sqrt{1 + \frac{\xi\delta}{\alpha}}$$

A plot of this universal solution is shown in figure 3, and a comparison with a helium transient experiment is shown in figure 4. Details are given in [6].

References [7, 8, and 9] give transient turbulent heat transfer analyses which have been applied to data with fluids at room temperature and above.

For a compressible fluid, the effect of thermal expansion in the heated boundary layer should be considered. In a closely confined channel this expansion, or attempted expansion, may cause a local pressure rise and consequent axial pressure gradient, as described in the next section. In a more open channel, and/or on a shorter time scale, this expansion has the effect of pushing the colder fluid away from the heated surface, thus decreasing the effective heat transfer rate from that predicted for thermal diffusion into an incompressible fluid.

The magnitude of this thermal expansion effect can be estimated from the one-dimensional equations for mass, momentum, and energy conservation in the fluid. Neglecting curvature of the heat transfer surface, one finds that the thermally induced velocity approaches an asymptotic limit as a function of distance from the surface. For an ideal gas, this asymptotic velocity limit is

$$v = \frac{(\gamma-1)}{\gamma}\frac{q}{P} :$$

where γ is the heat capacity ratio, q is the wall heat flux, and P is the pressure. A somewhat larger limit is obtained for a near-critical fluid where both the thermal expansivity and the compressibility are larger. The corresponding corrections to the wall temperature as a function of time, for several values of heat flux, are shown in figure 5 for a near-critical fluid (helium at 5.2 K and 0.25 MPa). One can see the deviation in wall temperature from that of the incompressible fluid model is almost uniquely determined by the calculated temperature rise, and in this sense is almost independent of heat flux or time. For temperature rises above about 1 kelvin, the error in the wall temperature rise predicted by the incompressible fluid model is about 50 percent in this case. Inclusion of axial flow, for example due to the onset of natural convection, may result in further significant effects.

Transient Heat Transfer into a Coolant Channel

In the practical case of a coolant channel connected at both ends to pressure or temperature reservoirs, or other devices, additional factors enter into the dynamic response to transient heat input.

Consider the practical case of transient heat input into a coolant channel of high L/D ratio, as illustrated in figure 6. The transient heat input may represent some off-design situation. Assume that the coolant channel is connected by valves (or other flow impedances) at either end to constant pressure reservoirs. Various systems can be modeled by changing the flow impedances of the valves. The heat input can be either localized or distributed uniformly along the channel length.

As the heat input develops, local boiling or compressible fluid phenomena as described above may occur. In addition, the local pressure will begin rising as the first step in perturbing the system flow rate. Thus an additional pressure gradient will be developed between the site of the heat input and the ends of the flow channel where the pressure generally is more nearly constant (depending on the flow impedances of the valves). This additional pressure gradient will cause a perturbation in the axial fluid velocity, and hence in the heat transfer rate. For a high L/D ratio this velocity perturbation may be quite significant. An example is seen in figure 7, calculated for a practical helium-cooled superconducting cable where 1 meter of cable is subjected to a pulsed heat input. Induced velocities corresponding to Reynolds numbers of up to 10^5 are seen, and the region of perturbed fluid velocities spreads away from the site of the initial perturbation with sonic velocity. A similar situation may exist in certain hypothetical failures of nuclear reactor cooling systems. Kawamura [8] has shown that the effective, time dependent heat transfer rate may differ by up to an order of magnitude from the (quasi static) value calculated from the Dittus-Boelter correlation using the time-dependent fluid velocity. We have an experiment under way to explore this question, but results are not available at this time.

Computer Modeling

Most of these studies have been supported by numerical modeling of the processes. For compressible fluids, the modeling requires hyperbolic differential equations specifying mass, momentum, and energy balance in a fluid with a finite sound velocity, v_{sound}. Numerical integration of hyperbolic differential equations often encounters problems in the correct specification of boundary conditions. Numerical stability is guaranteed only when certain combinations of thermodynamic parameters are specified at the boundaries, as determined from the characteristic velocities v and v ± c, for the system. It turns out that these theoretically derived thermodynamic combinations often do not relate easily to the practical constraints in the problem. For example, when using P, T, v (pressure, temperature, and velocity) as the independent parameters in the differential equations, stability is guaranteed only when one specifies boundary conditions. Numerical stability is guaranteed only when certain combinations of thermodynamic parameters are specified at the boundaries, as determined from the characteristic velocities v and v ± v_{sound}, for the system. It turns out that these theoretically derived thermodynamic combinations often do not relate easily to the practical constraints in the problem. For example, when using P, T, v (pressure, temperature, and velocity) as the independent parameters in the differential equations, stability is guaranteed only when one specifies

$$\left(\frac{P}{\rho}\frac{\partial\rho}{\partial P}\right)_s \pm \frac{v}{v_{sound}}$$

and

$$T\left(1 - \left(\frac{P}{T}\frac{\partial T}{\partial P}\right)_s\right)$$

at the boundaries. In practice, numerical stability can be obtained for at least some problems when one simply defines, for example, P, T, and v at the boundaries. The state of knowledge of hyperbolic equations is not sufficient to provide good guidelines about these boundary specifications.

We have found that ρ, T, v coordinates often are unusable because of this stability problem. P, T, v coordinates are a definite improvement, but still marginal. P, H (enthalpy), v coordinates generally are even better. It is possible that other groups of coordinates might result in further improvement in stability. Further studies in this direction are underway.

References

1. Steward, W. G., "Transient Helium Heat Transfer Phase I - Static Coolant", Int. J. Heat Mass Transfer, Vol. 21, pp. 863-874 (1978).

2. Giarratano, P. J., and Frederick, H. V., "Transient Pool Boiling of Liquid Helium Using a Temperature-Controlled Heater Surface", Adv. Cry. Eng., Vol. 25, pp. 455-466 (1980).

3. Giarratano, P. J., Lloyd, F. L., Mullen, L. O., and Chen, G. B., "A thin platinum film for transient heat transfer studies", Temperature--Its Measurement and Control in Science and Industry, Vol. 5, pp. 859-863 (1982) American Inst. Phys., NY, NY.

4. Carslaw, H. S., and Jaeger, J. C., "Conduction of Heat in Solids", Oxford Clarendon Press (1947).

5. Schlichting, H., "Boundary Layer Theory", Fourth Edition, Ch. XIX, McGraw-Hill Book Company, Inc., (1960).

6. Steward, W. G., "Transient Turbulent Heat Transfer", to be published.

7. Wang, C. C., Chung, B. T. F., and L. C. Thomas, "Transient Heat Transfer for Turbulent Boundary Layer Flows with Effects of Wall Capacity and Resistance", Sixth International Heat Transfer Conference, Toronto, Canada, 1978.

8. Kawamura, Hiroshi, "Experimental and Analytical Study of Transient Heat Transfer for Turbulent Flow in a Circular Tube", International Journal of Heat and Mass Transfer, Vol. 20, (1977).

9. Kawamura, Hiroshi, "Analysis of Transient Turbulent Heat Transfer in an Annulus: Part II: Heating Element with Heat Capacity and Thermal Resistance", translated from JSME, Vol. 42, No. 356, 1976.

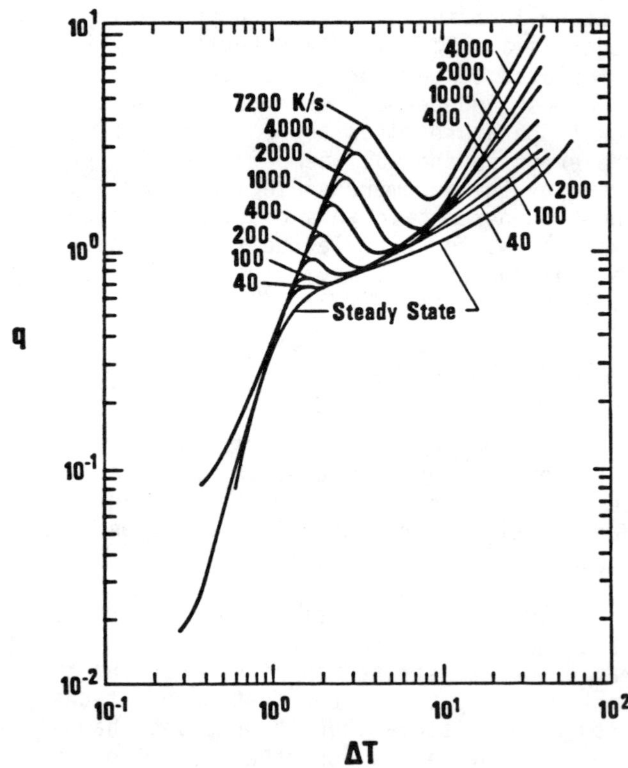

Figure 1. Helium boiling curve for various constant rates of temperature rise obtained with temperature controlled carbon film heater surface.

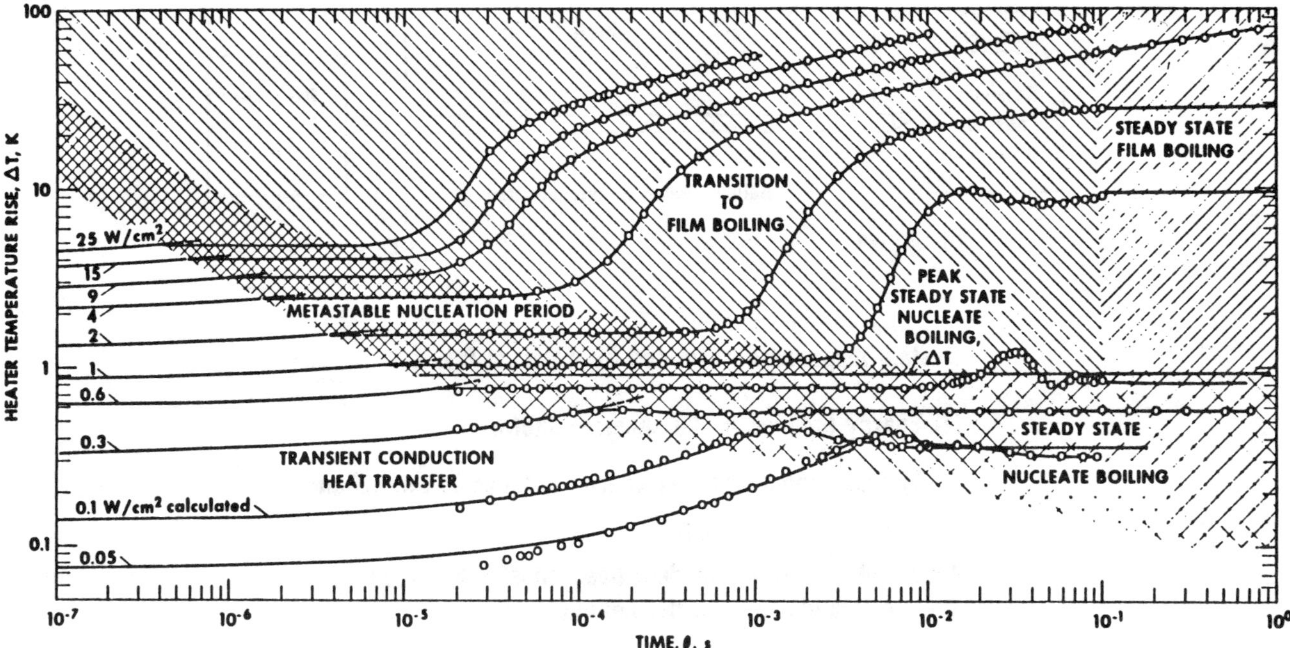

Figure 2. Transient heat transfer from a vertical carbon heater to saturated liquid helium at 4 K.

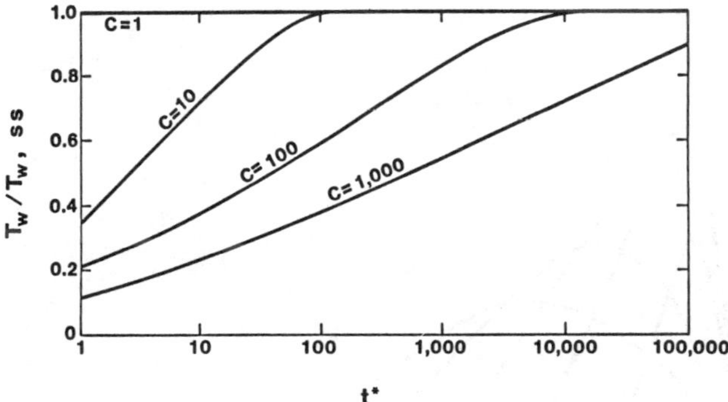

Figure 3. Ratio of transient-to-steady state wall temperature for turbulent convection from a flat heated surface.

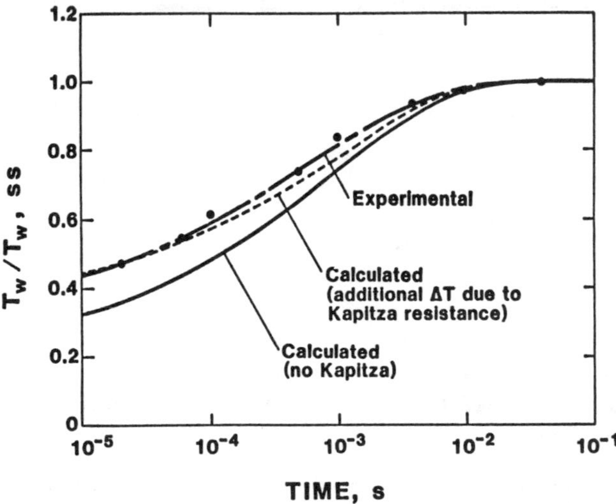

Figure 4. Transient turbulent convection—calculation compared with experiment.

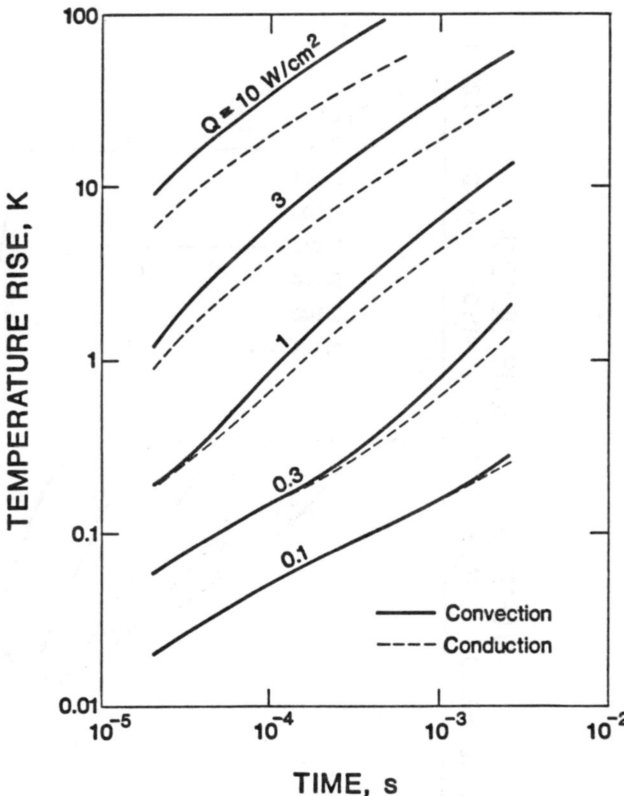

Figure 5. Comparison of transient transverse convection with transient conduction for near critical helium. For these numerical results the initial conditions are 0.25 MPa and 5.2 K. The critical pressure and temperature for helium are 0.25 MPa and 5.2 K.

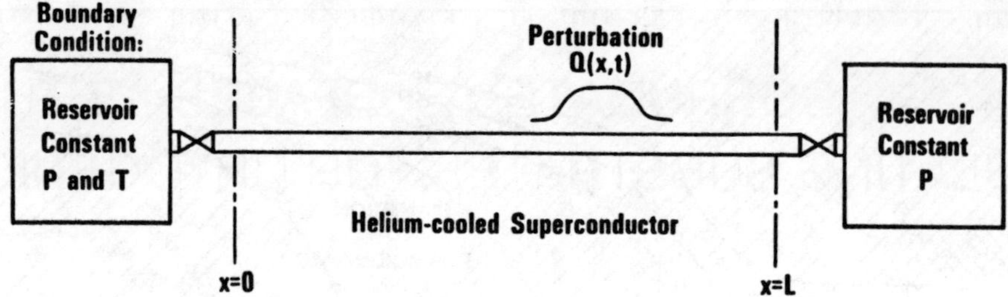

System response determined by 4 partial D.E.'s

(1) Energy balance in the superconductor

(2-4) Mass, momentum, & energy balance in the helium

with

time and velocity-dependent heat transfer between the superconductor and the helium

Figure 6. Schematic for transient heat transfer in high L/D systems in which induced axial velocity may be significant.

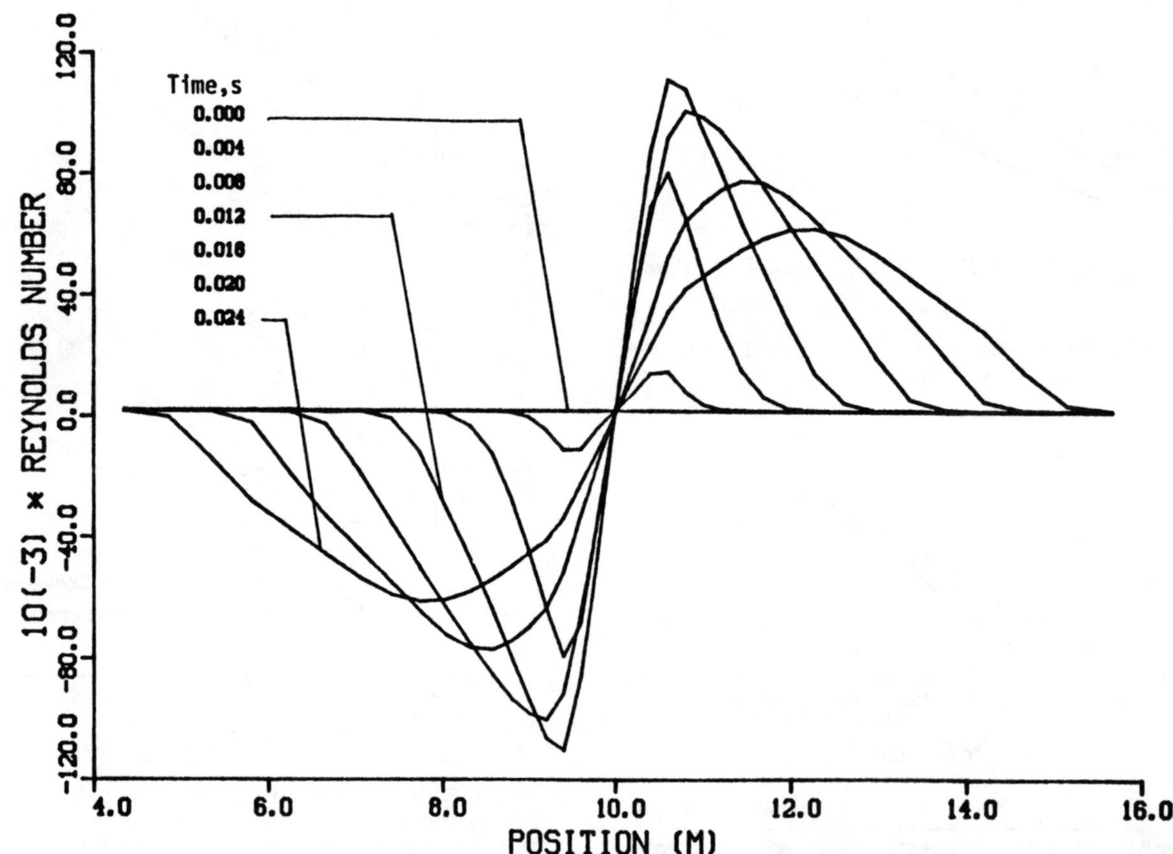

Figure 7. Heat induced Reynold's number in supercritical helium computed as a function of position and time following an initial heat input of 60 J/kg of conductor. The heat is applied to a 1m section centered in a 20 m conductor.

CORRUGATED SURFACES AS THERMALLY INSULATING SUBSTRATES FOR CRYOGENIC LIQUIDS

IFIYENIA KECECIOGLU

Westinghouse Research & Development Center
1310 Beulah Road
Pittsburgh, PA 15235

and

ROBERT C. REID

Department of Chemical Engineering
Massachusetts Institute of Technology
Cambridge, MA 02139

The effectiveness of corrugated surfaces, overlying thermally conductive substrates, in limiting the rate of evaporation of cryogenic liquids is analyzed.

A mathematical model that incorporates the different processes of heat transfer through the contact and air space regions between the corrugation and the substrate, and also the time-dependent heat conduction within the substrate, has been developed. The overall thermal resistance between the distant regions of the substrate and the corrugation has been interpreted as due to the series combination of an average uniform surface thermal resistance, a thermal resistance corresponding to the propagation of a one-dimensional temperature front within the substrate, and a constriction thermal resistance caused by the nonuniformity of the heat transfer coefficient at the corrugation substrate interface. An analysis of available experimental data shows that the constriction resistance significantly enhances the thermal insulation provided by a corrugated surface.

Liquefied natural gas (LNG) is a volatile liquid mixture of methane, some ethane, propane, and traces of the heavier hydrocarbons. At atmospheric pressure it has a boiling temperature of approximately -158°C. Consequently, should an accident occur at a storage facility and a large spill of LNG take place, the LNG would boil on contact with the ground and form a hydrocarbon vapor cloud. When mixed with air, this cloud would form a flammable mixture if the hydrocarbon concentration were in the range of 5 to 15% by volume.

The increasingly urgent and growing demand for energy has fueled the need to import and store vast quantities of LNG. The safety hazard posed by the operation of large LNG storage facilities in the proximity of urban population centers has led to government regulations designed to reduce the rate of boil-off of the LNG and to contain the vapor in the vicinity of the original spill. The NFPA regulations, in particular, specify that LNG storage facilities be diked in such a manner as to contain the entire liquid content of the full storage tank and that thermally insulating materials be used in their construction.

The present study was made in order to understand and characterize the effective thermal insulation provided by corrugated surfaces when placed in contact with soil (1) and to explain the results of experiments with LNG spills on corrugated aluminum conducted at the MIT LNG Research Center (2).

THEORETICAL ANALYSIS OF CRYOGEN BOIL-OFF RATE

The rate at which the liquid cryogen boils off is determined by the rate at which heat arrives at the cryogen-corrugation interface from regions located within the semi-infinite substrate. It is therefore controlled by the overall thermal resistance of heat flow that intervenes between the constant boiling temperature of the liquid cryogen and the constant and uniform temperature deep within the substrate.

The analysis is presented in two parts. The first part deals with the heat flow within a conductive substrate for a specified spatial distribution of the surface heat transfer coefficient. The second part deals with the problem of quantifying the surface heat transfer coefficient in the contact and air-space regions between the corrugation and the substrate. The problem of determining the extent of the contact region for a specified load is also addressed here. The theoretical models and the experiments which were performed to validate them are described in greater detail by Kececioglu (1) and will shortly appear as a journal publication.

Substrate Heat Transfer: Theoretical Model and Assumptions

The geometry of the corrugated surface in contact with the substrate is shown in Figure 1. Its undeformed nominal profile is assumed to be given by:

$$d(x) = a(1-\cos\frac{2\pi x}{\lambda}) \ . \qquad (1)$$

The heat flow within a conductive substrate for a specified spatial distribution of the surface heat transfer coefficient is obtained by solving the unsteady heat conduction equation for a homogeneous, isotropic substrate with constant physical properties. Then, the governing equation for the nondimensional local substrate temperature, $\Theta \equiv (T_s - T)/(T_s - T_L)$, is given by:

$$\frac{\partial^2\Theta}{\partial\xi^2} + \frac{\partial^2\Theta}{\partial\eta^2} = \frac{\partial\Theta}{\partial t^*} \ . \qquad (2)$$

Equation (2) is solved subject to the following initial and boundary conditions:

Initial condition:

$$\Theta\ (\xi,\ \eta,\ 0) = 0 \qquad (3)$$

Boundary conditions:

$$\Theta\ (\xi,\infty,t^*) = 0\ ;\ \left.\frac{\partial\Theta}{\partial\xi}\right|_{\xi=0} = 0\ ; \qquad (4,5)$$

$$\left.\frac{\partial\Theta}{\partial\xi}\right|_{\xi=1} = 0 \qquad (6)$$

$$\frac{q(\xi,t^*)}{h_a\ (T_s - T_L)} = -\frac{1}{Bi}\left.\frac{\partial\Theta}{\partial\eta}\right|_{\eta=0}$$

$$= f(\xi)\ (1-\Theta\left.\right|_{\eta=0}) \ . \qquad (7)$$

The exact solution to the above governing equations can be obtained numerically. However, we seek here a simpler but physically meaningful solution. Accordingly, we replace the boundary condition of Equation (7) by the approximate boundary condition given by Equation (8). That is, we assert that the ratio of the heat flux at the surface to its average value is equal to the ratio of the heat transfer coefficient to its average value; this means that the heat flux and the surface heat transfer coefficient vary proportionately. Using Equation (7) in Equation (8), we obtain Equation (9):

$$\frac{q(\xi,t^*)}{q_a(t^*)} \simeq \frac{h(\xi)}{h_a}\ ; \qquad (8)$$

$$\left.\frac{\partial\Theta}{\partial\eta}\right|_{\eta=0} = -\frac{Bi\ q_a(t^*)}{h_a\ (T_s - T_L)} \cdot f(\xi)\ . \qquad (9)$$

From Equation (7) and the definition for h_a, we have:

$$\frac{q_a(t^*)}{h_a(T_s - T_L)} = 1 - \int_0^1 f(\xi)\ \Theta\left.\right|_{\eta=0} d\xi . \qquad (10)$$

The solution of Equation (2), subject to the initial condition (3) and boundary conditions (4, 5, 6, and 9), can be obtained in Laplace transform space by the separation of variables technique. This is given below:

$$\bar{\Theta}(\xi,\eta,s)$$

$$= \sum_{m=0}^{\infty} E_m \cos\pi m\xi\ e^{-\sqrt{s+(\pi m)^2}\ \eta}, \qquad (11)$$

where

$$E_m = \frac{2Bi\ \bar{q}_a(s)}{h_a(T_s - T_L)} \cdot \frac{f_m}{\sqrt{s+(\pi m)^2}} \qquad (12)$$

$$f_m = \int_0^1 f(\xi)\ \cos\pi m\xi\ d\xi\ (m > 0)\ ;$$

$$f_o = 1/2\ . \qquad (13)$$

Hence, from Expression (10) we obtain for heat flux in Laplace transform space:

$$\bar{q}_a(s) = h_a(T_s - T_L)$$

$$\cdot\left[\ s(1 + \frac{Bi}{\sqrt{s}} + 2Bi\sum_{m=1}^{\infty}\frac{f_m^{\ 2}}{\sqrt{s+(\pi m)^2}})\right]^{-1} \ . \qquad (14)$$

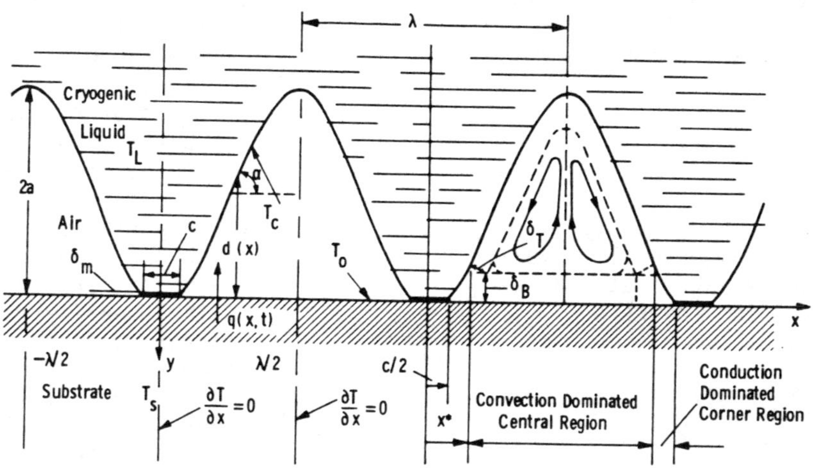

Figure 1. Cryogenic liquid spilled on a corrugated surface in contact with a semi-infinite substrate.

For the purpose of interpretation, we now define a time-dependent overall heat transfer coefficient, $H_a(t^*)$, by:

$$q_a(t^*) \equiv H_a(t^*)\ (T_s - T_L)\ , \qquad (15)$$

where in Laplace transform space,

$$\frac{1}{\bar{H}_a(s)} = s\ [\ \frac{1}{h_a} + \frac{Bi}{h_a\sqrt{s}} + \frac{2Bi}{h_a} \sum_{m=1}^{\infty} \frac{f_m^{\ 2}}{\sqrt{s+(\pi m)^2}}\ ]\ . \qquad (16)$$

The result of Equation (16) can be interpreted as the statement that the overall resistance to heat transfer between the temperatures T_s and T_L is the sum of three separately identifiable thermal resistances. The first of these, given by the first term of Equation (16), is the resistance to heat transfer at the surface. In the case of an isothermal substrate (i.e., $k_s \to \infty$ or equivalently $Bi \to 0$) the second term in (16) is equal to zero. Hence, we conclude that the second term represents the resistance to heat transfer offered by a growing region of the substrate lying between the surface temperature and the temperature front propagating away from it when the surface heat transfer coefficient is uniform and equal in value to h_a. The last term of Equation (16) represents the additional thermal resistance due to the nonuniformity of the heat transfer coefficient at the surface. Such nonuniformities force the heat flow streamlines to curve towards the regions of lower thermal surface resistance.

In the present study we assume a stepwise distribution in the surface heat transfer coefficient with different, but uniform, values over the contact and air-space regions. Then,

$$h(x) = h_1 \qquad 0 \leq x \leq c/2$$

$$h(x) = h_2 \qquad c/2 \leq x \leq \lambda/2. \qquad (17)$$

Hence,

$$h_a = \phi h_1 + (1 - \phi)h_2\ ;\ \phi \equiv c/\lambda. \qquad (18a;b)$$

Then,

$$f_m = (\frac{h_1 - h_2}{h_a})(\frac{\sin m\pi\phi}{m\pi}),\ m \neq 0;\ f_o = 1/2. \qquad (19)$$

If the heat transfer coefficient is uniform over the surface of the substrate, then $h_1 = h_2$, $f_m = 0$, and we recover the solution for uniform heat flow.

The mass of cryogen boiled off after a time, t, is given by:

$$m(t) = \int_0^t \frac{q_a(t)}{\Delta H_V}\ dt\ , \qquad (20)$$

where ΔH_V is the latent heat of vaporization of the cryogen. Alternatively, it can be expressed in Laplace transform space by:

$$\bar{m}(s) = \frac{1}{\Delta H_V} \cdot \frac{1}{s} \cdot \bar{q}_a(s) \; , \qquad (21)$$

where $\bar{q}_a(s)$ is given by Equation (14). These expressions were numerically inverted exactly to yield the average heat flux and mass boiled off using the inversion algorithm of Stehfest (3).

Surface Heat Transfer Coefficient: Theoretical Model and Assumptions

Theoretical expressions were developed for 1) the heat transfer coefficient h_1, in the region of contact between the valleys of the corrugation and the plane surface of the substrate; and 2) the heat transfer coefficient h_2, in the air-space regions between the peaks of the corrugation and the plane surface of the substrate. A brief description of the theoretical model is given below.

Heat Transfer Coefficient h_1 in the Contact Region. The contact area is viewed as a region of contact between two rough surfaces. Then, the transfer of heat through the contact region as a whole depends upon the heat conducted across each contact between asperities on the opposing rough surfaces (h_{1S}) and upon conduction (h_{1C}) and radiation (h_{1R}) of heat across the intervening interasperity air gaps. The total heat flux across the macroscopic contact is equal to the sum of the heat fluxes across the individual asperity contacts and air gaps.

The expressions derived for the contact heat transfer coefficient, h_1, are summarized in Table 1. A detailed description of the derivation of these expressions is given by Kececioglu (1). The assumptions made in deriving them are summarized below. They are:

- The contact length c (Figure 1) was established according to a Hertzian theory of contact between smooth, nonconforming, elastic bodies.

- The effective stiffness of the corrugation, E_{2eff}, which is much less than that of the corrugation material, was obtained by multiplying the Young's modulus of the corrugation material (E_2) by the ratio of the cross-sectional area of a wavelength of the corrugation to the cross-sectional area of a cylinder of radius R; R is the radius of curvature of the corrugation at the point of contact.

- The asperities of the contacting rough surfaces were characterized by a Gaussian probability distribution of the asperity heights with a standard deviation, σ, and a mean asperity slope, m. As a result, the contact area dimensions (a_i, b_i) (see Figure 2) were related to the mean asperity separation, Y, of the rough surfaces through relations developed by Cooper, Mikic, and Yovanovich (4). Then,

$$\rho^2 \equiv \frac{\Sigma \pi a_i^2}{A_c} = \frac{1}{2}\,\mathrm{erfc}\left(\frac{Y}{\sigma\sqrt{2}}\right) \qquad (22)$$

$$A_c \simeq \Sigma\, \pi b_i^2 \; ; \quad Y \simeq \delta_m \qquad (23)$$

and

$$\frac{2\Sigma a_i}{A_c} = \frac{m}{2\sigma\,\sqrt{2\pi}}\, \exp\left(\frac{-Y^2}{2\,\sigma^2}\right) \; , \qquad (24)$$

where ρ^2 is the ratio of asperity contact area to the base area of the asperities and δ_m is the mean separation distance between contacting surfaces (see Figures 1 and 2).

- Locally steady-state conditions were assumed to apply to the transfer of heat through the interasperity fluid and the asperity contacts.

Heat Transfer Coefficient h_2 within the Airspaces. The approximate theoretical expressions derived for the heat transfer coefficient, h_2, in the air space between the corrugation and the substrate are included in Table 1. These expressions take into account the transfer of heat due to thermal radiation and also due to conduction and natural convection through the air space. If the adverse thermal gradients induce buoyancy forces capable of overcoming viscous friction within the air space, then the conduction through the fluid will be enhanced by natural convection. The gravitational equilibrium of the fluid

Table 1

EQUATIONS USED IN COMPUTATION OF THE AVERAGE SURFACE HEAT TRANSFER COEFFICIENT h_a[(a)]

Heat Transfer Coefficient h_1 within the Contact Region

$$h_1 = h_{1s} + h_{1C} + h_{1R}$$

Asperity Conduction:

$$h_{1s} = k_s(1-\rho)^{-\frac{3}{2}} \frac{2\Sigma a_i}{A_c}$$

Interasperity Fluid Conduction:

$$h_{1C} = \frac{k_f}{\delta_m}(1-\rho^2)$$

Radiation:

$$h_{1R} = 4\,\bar{\sigma}\,\varepsilon\,T_m^3\,f_{12}$$

Heat Transfer Coefficient h_2 within the Air Spaces

$$h_2 = h_{2C} + h_{2R}$$

Conduction and Convection:

$$h_{2C} = \phi^{*(b)}\,h_{2C}^* + (1-\phi^*)h_{2NC}^*$$

Conduction in Corner Region:

$$h_{2C}^* = \frac{k_f}{a}\,\frac{1}{\pi\phi^*(1-\phi)}\int_{\pi\phi}^{\zeta^{*(c)}} \frac{a}{d(\zeta)}\,d\zeta$$

$$d(\zeta) \simeq \delta_m + \frac{1}{2}a\left(\zeta^2 - (\pi\phi)^2\right)\ ;\ (\zeta,\pi\phi) \ll 1.$$

Natural Convection:

$$h_{2NC}^* = 0.157\,\frac{k_f}{a}\,\frac{2}{1+(\cos\alpha)^{0.715}}\,Ra^{0.285}\ ;\alpha \leq 60^o$$

Radiation:

$$h_{2R} = 4\,\bar{\sigma}\,\varepsilon\,T_m^3$$

(a) $h_a = \phi h_1 + (1-\phi)h_2;\ \phi \equiv c/\lambda = \frac{2}{\pi}\sqrt{\frac{P_{APP}}{E'}\frac{\lambda}{\pi a}}\ ;\ \frac{1}{E'} = \frac{1-\nu_1^2}{E_1} + \frac{1-\nu_2^2}{E_{2eff}}$.

(b) $\phi^* \equiv \frac{1}{\pi}\ \frac{\zeta^*-\pi\phi}{1-\phi}$

(c) $\zeta^* \simeq \sqrt{\frac{2(\delta_B+\delta_T\cos\alpha)}{a}} = \sqrt{\frac{2k_f}{h_{2NC}^*\,a}}$

subjected to an adverse temperature gradient becomes unstable when the Rayleigh number, Ra, exceeds a certain critical value, Ra_c (for plane inclined layers, $Ra_c.\cos\alpha \leq 1708$, as given by Buchberg et al. (5) where $\alpha \leq 60^o$ is the angle of inclination to the horizontal).

The geometry of the air space enclosed between the corrugation and the substrate can be viewed, approximately, as an isosceles triangle formed from two inclined surfaces representing the corrugation and a horizontal surface representing the substrate. In a recent work Flack, Konopnicki, and Rooke (6) considered the heat trans-

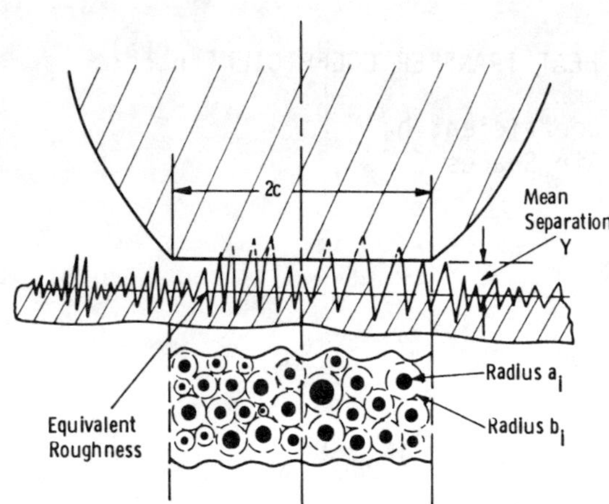

Figure 2. Idealized contact geometry with flat nominal surfaces and a single equivalent rough surface.

ferred between two inclined faces of an isosceles triangle with a horizontal adiabatic base. In this system the heated and the cooled surfaces have the same inclination to the horizontal, and the flow patterns are different from those that would occur in the present geometry. Two of the features of the present case, however, namely, the existence of thermal boundary layers at the hot and cold surfaces and the occurrence of a highly conductive corner region were observed by the above-mentioned authors.

In our analysis we identified two distinct regions lying within the airspace ($c/2 \leqslant x \leqslant \lambda/2$). The first of these regions ($c/2 \leqslant x \leqslant x^*$) is located adjacent to the contact region. It contains a layer of essentially stagnant fluid through which heat is transferred by molecular diffusion. Its outer boundary at $x = x^*$ is taken to be fixed by the thickness of the thermal boundary layers, δ_T and δ_B at the top (i.e., inclined) and bottom (i.e., substrate) surfaces, respectively, as shown in Figure 1. As justified by Kececioglu (1), the temperature profile in this corner region between the substrate temperature and the liquid cryogen temperature is assumed to be linear.

The second region, within which convective transport of heat occurs, extends over $x^* \leqslant x \leqslant \lambda/2$. Within this region we idealize the situation and assume the existence of uniform effective thermal boundary layers of thickness δ_T and δ_B at the top and bottom surfaces, respectively. Within these thermal boundary layers the heat transport normal to the surface is assumed to be primarily by conduction. The temperature in the region lying between the two boundary layers is assumed to be uniform and equal to T_m^*. The heat transfer coefficient in this region is derived by idealizing the inclined surface as a plane of length $(\lambda/2 - x^*)/\cos\alpha$ and of inclination α to the horizontal, where $\tan \alpha = 2\pi a/\lambda$ is the slope at the midpoint of the corrugation (i.e., $x = \lambda/4$), and by using the correlations of Buchberg et al. (5) for natural convection over plane inclined layers with angles of inclination to the horizontal as large as 60°. In these correlations the extensive results for plane horizontal layers were scaled to the inclined layer case by replacing the acceleration due to gravity, g, by its resolved component $g.\cos\alpha$(7,8). For the experiments reported here the modified Rayleigh number, $Ra.\cos\alpha$, was in the range of $2500 \leqslant Ra.\cos\alpha \leqslant 10^6$, lying within the range of validity of the expressions for the heat transfer coefficients given in Table 1. The results of experiments performed to validate these approximate heat transfer models were previously reported (1).

COMPARISON OF THEORETICAL AND EXPERIMENTAL BOIL-OFF RATES FOR CORRUGATIONS IN CONTACT WITH SOILS

The prediction of the present theoretical model for the mass of cryogen boiled off was tested through comparison against the experimental data for liquid methane boil-off on corrugated surfaces. These experiments, which were previously reported by the MIT LNG Research Center (9), were conducted on corrugated surfaces laid over layers of soil packed within a 5-cm thick styrofoam box (See Figure 3). The rate at which a measured mass of liquid methane, spilled suddenly onto the corrugation, boiled off was continuously measured by recording the weight of the entire apparatus. The temperature of the cryogen and its vapor and the temperature of the soil at different depths from the contact surface were also measured. No provision was made to control or measure the total contact load. The initial tamping of the corrugation on the soil, however, was varied in an effort to encompass what was believed to be the practical range of contact area between the soil and the corrugation. The experimental results are reproduced in Figure 4, and the dimensions of the corrugation are given in Table 2.

These experimental results for the mass of liquid methane boiled off as a function

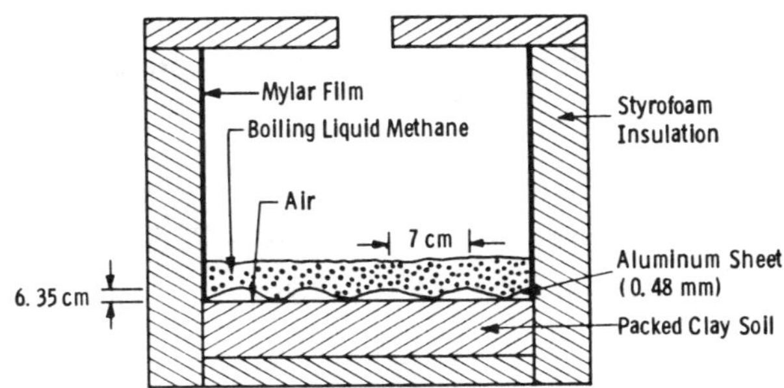

Figure 3. The experimental set-up for the measurement of the boil-off rate of liquid methane on a corrugation in contact with clayey soil.

of time were matched using the previously described theoretical models for the transient substrate heat transfer and heat transfer at the contact surface. The heat transfer coefficients h_1 and h_2 in the contact and air-space regions and the contact area fraction, ϕ, as a function of the contact load were computed from the expressions given in Table 1. These values, in turn, were used to compute the average heat transfer coefficient, h_a (Equations 18a, b), the Fourier coefficients, f_m (Equation 19), and the Biot number, Bi, required to evaluate the Laplace transform of the mass boiled off from Equation (21). The mass boiled off as a function of time was obtained by numerically inverting this expression using the inversion algorithm of Stehfest (3).

At the very outset we estimated that for the nucleate boiling regime present in these experiments, the boiling heat transfer coefficient for liquid methane was large, and its effect could be neglected. The mean interasperity separation distance, δ_m, was assumed to be equal to the mean grain size, and the standard deviation of asperity heights, σ, was assumed to be equal to four times the mean grain size on the basis of grain size distributions reported in the literature for typical clayey soils. For the East Coast clayey soil used, the following material parameters were estimated from the literature (10-12):

$\delta_m = 2.5 \times 10^{-5}$ m; $\sigma = 4\ \delta m$
$k_s = 0.4$ W/mK ; $\alpha_s = 0.77 \times 10^{-6}\ m^2/s$
$E_2 = 0.5173 \times 10^8\ N/m^2; \nu_2 = 0.3.$

For the above values of the soil parameters, the range of tamping used in the experiments was simulated by adjusting the apparent contact pressure, P_{APP}, until the boil-off curves matched the experimentally measured boil-off curves. The initial temperature of the soil was taken to be the average (15.5°C) of the values reported. The theoretically computed boil-off curves for a low apparent contact pressure of 0.3277 x $10^4 N/m^2$ and a high apparent contact pressure of 0.450 x $10^5\ N/m^2$ are shown in Figure 4. The agreement between the computed and measured curves for these realistic apparent contact pressures is very good. In this figure the boil-off curve that would be obtained for a uniform heat transfer coefficient equal to the average surface heat transfer coefficient, h_a, but without the constriction thermal resistance within the substrate, is also shown. A very significant reduction in boil-off results from this effect. A few relevant results of these computations are given below.

Quantity	Low Load	High Load
P_{APP} (N/m^2)	0.3277×10^4	0.4500×10^5
ϕ	0.0520	0.1929
h_1 $(W/m^2\ K)$	1071.3	1071.3
h_2 $(W/m^2\ K)$	23.173	15.578
h_a $(W/m^2\ K)$	77.74	219.0
$h_1\phi/h_a$	0.717	0.944
$h_2(1-\phi)/h_a$	0.283	0.056

An important observation that can be made from these results is that the heat transfer through the air-space region at low load is a significant fraction of the total heat transfer and cannot be neglected; at high loads, however, the contact area heat transfer becomes dominant. An additional advantage of using corrugations in the present application is that the increased extraction of heat through seepage of the cryogen into the soil is prevented. The experimental boil-off curve of liquid methane for a direct spill on a packed soil substrate (2) was correlated by the equation $m(t) = t^{1/2}$.

CONCLUSIONS

A mathematical model of heat transfer between a corrugated surface and an underlying thermally conductive substrate was developed to predict the rate of boil-off of cryogen spilled on the corrugated surface. The overall thermal resistance between the distant regions of the substrate and the corrugation was shown to be significantly increased by a thermal constriction resistance introduced by the nonuniformity of the heat transfer coefficient at the corrugation-substrate interface. Analysis of available

Table 2

THE GEOMETRY AND EFFECTIVE STIFFNESSES OF THE ALUMINUM CORRUGATION USED IN THE EXPERIMENTS

No. of Contacts	a CM	λ CM	t CM	R CM	a/λ	t/a	α Deg.	$\frac{E_{2eff}}{E_2}$	$E' \times 10^{10}$ N/m^2
4	0.6350	7.000	0.0480	1.9546	0.0907	0.0756	29.68	0.0302	0.222

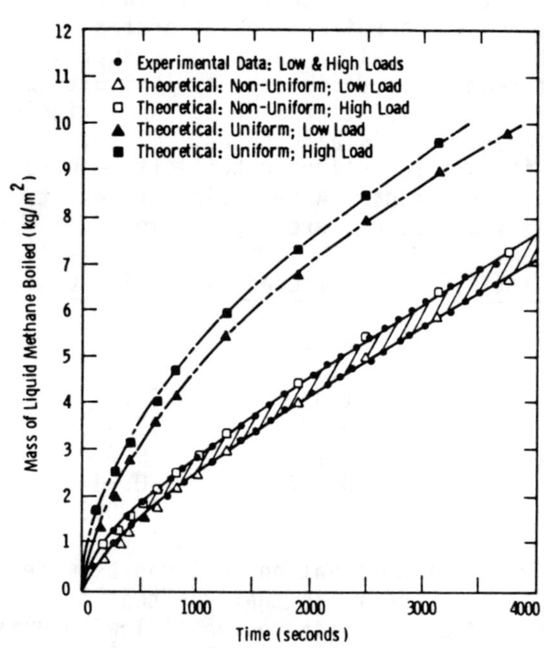

Figure 4. The mass of liquid methane boiled off from the corrugated aluminum surface in contact with a packed soil substrate.

experimental boil-off data showed that the constriction resistance confers upon the corrugation a significantly greater degree of thermal insulation than that provided by a uniform layer of insulating material with the same average steady-state heat transfer coefficient. The present model provides a systematic means of predicting the boil-off rates on corrugations and of optimizing the corrugation geometry for maximum thermal insulation under specified material properties and loading conditions.

ACKNOWLEDGMENTS

The financial support provided for this research by the Gas Research Institute is gratefully acknowledged. The assistance provided by the Chemical Engineering Department of the Westinghouse Research and Development Center was invaluable in bringing this paper to publication.

NOTATION

A_c = macroscopic contact area (Eq. 23a; See Fig. 2).

a = amplitude of corrugation

a_i = contact radius (See Fig. 2)

Bi = Biot number ($Bi \equiv \frac{h_a \lambda}{2 k_s}$)

b_i = radius of adiabatic contact heat channel (See Fig. 2)

c = contact length of contact area per unit length of corrugation over one wavelength of corrugation (Table 1; Fig. 1)

d(x) = undeformed "nominal" profile of corrugation (Eq. 1)

E_1, E_2 = Young's moduli of substrate and corrugation material

E_{2eff} = effective Young's modulus of the corrugated surface

$$\left(\frac{E_{2eff}}{E_2} = \frac{t}{a} \frac{4k^3}{\pi} \int_0^{\pi/2} \sqrt{1 + k^2 \sin^2\theta}\, d\theta \;;\quad k \equiv \frac{2\pi a}{\lambda}\right)$$

$f(\xi)$ = nondimensional heat transfer coefficient $[f(\xi) \equiv h(\xi)/h_a]$

f_m = Fourier coefficients given by Eq. (13) and Eq. (19)

f_{12} = thermal radiation view factor between surfaces 1 & 2

g = gravitational acceleration

$H_a(t^*)$ = overall heat transfer coefficient (Eq. 15)

$\bar{H}_a(s)$ = overall heat transfer coefficient in Laplace transform space (Eq. 16)

h_a = average heat transfer coefficient over the surface of the substrate

$$(h_a \equiv \frac{2}{\lambda} \int_0^{\lambda/2} h(x)\, dx$$

$$= \int_0^1 h(\xi)\, d\xi \text{ ; Eq. 18a)}$$

$h(x)$, $h(\xi)$ = heat transfer coefficient defined by:

$$q(x,t) \equiv h(x)\,[T(x,0,t) - T_L]$$

h_1 = surface heat transfer coefficient in macroscopic contact area (Eq. 17 and Eq. 18a)

h_{1C}, h_{1R}, h_{1s} = heat transfer coefficients for heat transferred by conduction through the interasperity fluid, by radiation, and by conduction through the physical contacts between the two contacting surfaces (See Table 1).

h_2 = heat transfer coefficient within the air space of the corrugation

h_{2C}, h_{2R} = heat transfer coefficients for heat transferred by conduction, convection, and radiation within the air spaces of the corrugation (See Table 1)

$h_{2C}{}^*$ = heat transfer coefficient for heat transferred by conduction in the corner region of the air pocket of the corrugation (See Table 1)

$h_{2NC}{}^*$ = heat transfer coefficient for heat transferred by natural convection through the air pocket of the corrugation

k_f = thermal conductivity of fluid between substrate and corrugation

k_s = thermal conductivity of substrate

m = mean slope of the asperities of the equivalent rough surface in contact with a nominally flat smooth surface ($m \simeq 0.1$)

$m(t)$ = mass of cryogen boiled off as a function of time

$\bar{m}(s)$ = mass of cryogen boiled off in Laplace transform space

P_{APP} = apparent contact pressure

$$= \frac{\text{Total contact force on corrugation}}{\text{Projected area of corrugation}}$$

$q_a(t^*)$ = average surface heat flux for semi-infinite substrate (Eq. 10)

$\bar{q}_a(s)$ = average surface heat flux in Laplace transform space (Eq. 14)

$q(\xi,t^*)$ = surface heat flux (Eq. 7)

Ra = Rayleigh No.;

$$Ra \equiv \frac{g\, a^3\, \beta_m\, (T_o - T_L)}{\nu_m{}^2} \cdot \frac{\nu_m}{\alpha_m}$$

Ra_c = critical Rayleigh No.; if $Ra > Ra_c$, convection cells become established; if $Ra < Ra_c$, heat transfer is entirely by conduction

s = Laplace transform variable

$T(x,y,t)$ = temperature at position (x,y) within the substrate at time t

T_L = boiling temperature of the liquid cryogen

T_m : $4T_m{}^3 \equiv (T_1{}^2 + T_2{}^2)(T_1 + T_2)$; T_1, T_2: average temperatures of opposing surfaces of the air gap

$$T_m{}^* = \frac{\dfrac{T_o}{\delta_B} + \dfrac{T_L}{\delta_T \cos\alpha}}{\dfrac{1}{\delta_B} + \dfrac{1}{\delta_T \cos\alpha}}$$

T_o = average surface temperature of substrate

T_s = initial temperature of substrate

t = time or corrugation thickness

t^* = dimensionless time; $t^* = t/(\lambda^2/4\alpha_s)$

x = position along surface of semi-infinite substrate (See Fig. 1)

x^* = size of conduction-dominated corner region in air space of corrugation (See Fig. 1 and Table 1)

Y = separation of the nominally flat mean surfaces in the region of contact (Eq. 22 and Eq. 23b)

y = position inside substrate away from the corrugation-substrate interface

α = angle of inclination of fluid layer to the horizontal

α_m = thermal diffusivity of fluid evaluated at $1/2(T_o + T_L)$

α_s = thermal diffusivity of substrate

β_m = thermal volume expansion coefficient of the fluid evaluated at an average temperature $1/2(T_o + T_L)$

ΔH_V = latent heat of vaporization of liquid cryogen

$\delta_{B,T}$ = thermal boundary layer thicknesses for plane and inclined fluid layers (See Fig. 1).

δ_m = mean interasperity air-gap distance in the contact region (Eq. 23) (See Figs. 1 and 2)

ε = radiative emissivity (See Table 1)

ζ, ζ^* = $2\pi x/\lambda$; $2\pi x^*/\lambda$

η = $2y/\lambda$

Θ = dimensionless local substrate temperature; $\Theta = (T_s - T)/T_s - T_L)$

$\bar{\Theta}$ = Θ in Laplace transform space;

$$\bar{\Theta}(\xi,\eta,s) \equiv \int_0^\infty e^{-st^*}\, \Theta(\xi,\eta,t^*)\, dt^*$$

λ = wavelength of corrugation (see Fig. 1)

$\nu_{1,2}$ = Poisson's ratio for substrate and corrugation

ν_m = viscous diffusivity of air in air space evaluated at $1/2(T_o + T_L)$

ξ = $2x/\lambda$

ρ^2 = ratio of asperity contact area to base area of asperities (Eq. 22; See Fig. 2)

σ = standard deviation of the asperities of an equivalent rough surface (See Fig. 2) in contact with a nominally flat smooth surface

$\bar{\sigma}$ = Stefan-Boltzman constant

ϕ = fraction of the total surface area occupied by the contact region; $\phi \equiv c/\lambda$

ϕ^* = fraction of surface area occupied by air space in which heat is transferred by conduction (See Fig. 1 and Table 1)

LITERATURE CITED

1. Kececioglu, I., "Pool Boiling of Liquid Nitrogen on Corrugated Surfaces," S. M. Thesis, Chemical Engineering, MIT (1980).

2. Reid, R. C., Wang, R., "The Boiling Rates of LNG on Typical Dike Floor Materials," Cryogenics, 401, (1978).

3. Stehfest, H., "Algorithm 368; Numerical Inversion of Laplace Transforms," D-5, Communication of the ACM, 13, (1), 47, (1970).

4. Cooper, M. G., B. B. Mikic, and M. M. Yovanovich, "Thermal Contact Conductance," Int. J. Heat and Mass Transfer, 12, 279, (1969).

5. Buchberg, H., I. Catton, and D. K. Edwards, "Natural Convection in Enclosed Spaces - A Review of Application to Solar Energy Collection," J. Heat Transfer, Trans. ASME, 98, 182, (1976).

6. Flack, R. R., T. T. Konopnicki, and J. H. Rooke, "The Measurement of Natural Convective Heat Transfer in Triangular Enclosures," Trans. ASME, J. of Heat Transfer, 101, 648 (1979).

7. Clever, R. M., "Finite Amplitude Longitudinal Convection Rolls in an Inclined Layer," J. of Heat Transfer, Trans. ASME, Series C, 95, 407, (1973).

8. Hollands, K. G. T., T. E. Unny, G. D. Raithby, and L. Konicek, "Free Convection Heat Transfer Across Inclined Air Layers," ASME Paper No. 75-HT55 (1975).

9. MIT LNG Research Center; Final Report to the Gas Research Institute, Task VI, "Boiling of LNG on Typical Dike Floor Materials," (1978).

10. Chamberlain, E. J., D. M. Cole, and T. C. Johnson, "The Resilient Response of Two Frozen and Thawed Soils," J. of Geotechnical Eng., Div. of ASCE, 257, (1979).

11. Kersten, M. S., Proc. Highway Research Board, Univ. of Minnesota, 28, 161 (1948).

12. Cloessner, J. J., and R. F. Baron, "Thermal Conductivity of Frozen Soils at Cryogenic Temperatures," Advances in Cryogenic Engineering, 13, 655 (1968).

KINETICS

4 Reaction Kinetics and Transfer Process
25 Reaction Kinetics and Unit Operations
72 Recent Advances in Kinetics
73 Kinetics and Catalysis
83 Recent Advances in Kinetics and Catalysis

MATHEMATICS

31 Advances in Computational and Mathematical Techniques in Chemical Engineering
37 Applied Mathematics in Chemical Engineering
42 Statistical and Numerical Methods in Chemical Engineering
50 Optimization Techniques

MINERALS

15 Mineral Engineering Techniques
20 Liquid Metals Technology—Part I
43 Recent Advances in Ferrous Metallurgy
85 Fossil Hydrocarbon and Mineral Processing
173 Fundamental Aspects of Hydrometallurgical Processes
180 Spinning Wire from Molten Metals
216 Processing of Energy and Metallic Minerals

PETROCHEMICALS

34 Petrochemicals and Petroleum Refining
49 Polymer Processing
127 Declining Domestic Reserves—Effect on Petroleum and Petrochemical Industry
135 The Petroleum/Petrochemical Industry and the Ecological Challenge
142 Optimum Use of World Petroleum
212 Interfacial Phenomena In Enhanced Oil Recovery

PETROLEUM PROCESSING

34 Petrochemicals and Petroleum Refining
54 Hydrocarbons from Oil Shale, Oil Sands, and Coal
98 Methanol Technology and Economics
103 C_4 Hydrocarbon Production and Distribution
127 Declining Domestic Reserves—Effect on Petroleum and Petrochemical Industry
135 The Petroleum/Petrochemical Industry and the Ecological Challenge
142 Optimum Use of World Petroleum
155 Oil Shale and Tar Sands

PHASE EQUILIBRIA

2 Pittsburgh and Houston
3 Minneapolis and Columbus
6 Collected Research Papers
81 Phase Equilibria and Related Properties
88 Phase Equilibria and Gas Mixtures Properties

PROCESS DYNAMICS

36 Process Dynamics and Control
46 Process Systems Engineering
55 Process Control and Applied Mathematics
159 Chemical Process Control
214 Selected Topics on Computer-Aided Process Design and Analysis

SEPARATION

91 Unusual Methods of Separation
120 Recent Advances in Separation Techniques
192 Recent Advances in Separation Techniques—II

SONICS

1 Ultrasonics—Two Symposia
109 Sonochemical Engineering

THERMODYNAMICS

7 Applied Thermodynamics
44 Thermodynamics
140 Thermodynamics—Data and Correlations

TRANSPORT PROPERTIES

16 Mass Transfer
56 Selected Topics in Transport Phenomena
77 Fundamental Research on Heat and Mass Transfer

MISCELLANEOUS

26 Chemical Engineering Education—Academic and Industrial
48 Chemical Engineering Reviews
70 Small-Scale Equipment for Chemical Engineering Laboratories
76 High Pressure Technology
112 Engineering, Chemistry, and Use of Plasma Reactors
125 Vacuum Technology at Low Temperatures
143 Standardization of Catalyst Test Methods
160 Continuous Polymerization Reactors
182 Biorheology
183 The Modern Undergraduate Laboratory Innovative Techniques
185 Electro organic Synthesis Technology
186 Plasma Chemical Processing
187 Chronic Replacement of Kidney Function
194 Hazardous Chemical—Spills and Waterborne Transportation
203 A Review of AIChE's Design Institute for Physical Property Data (DIPPR) and Worldwide Affiliated Activities
204 Tutorial Lectures in Electrochemical Engineering and Technology
206 Controlled Release Systems
224 Cryogenic Processes and Equipment 1982

MONOGRAPH SERIES

1 Reaction Kinetics by Olaf Hougen
2 Atomization and Spray Drying by W. R. Marshall, Jr.
3 The Manufacture of Nitric Acid by the Oxidation of Ammonia—The DuPont Pressure Process by Thomas H. Chilton
4 Experiences and Experiments with Process Dynamics by Joel O. Hougen
5 Present, Past, and Future Property Estimation Techniques by Robert C. Reid
6 Catalysts and Reactors by James Wei
7 The 'Calculated' Loss-of-Coolant Accident by L. J. Ybarrondo, C. W. Solbrig, H. S. Isbin
8 Understanding and Conceiving Chemical Process by C. Judson King
9 Ecosystem Technology:Theory and Practice by Aaron J. Teller
10 Fundamentals of Fire and Explosion by Daniel R. Stull
11 Lumps, Models and Kinetics in Practice by Vern W. Weekman, Jr.
12 Lectures in Atmospheric Chemistry by John H. Seinfeld
13 Advanced Process Engineering by James R. Fair
14 Synfuels from Coal by Bernard S. Lee

SYMPOSIUM SERIES

ADSORPTION

74 Physical Adsorption Processes and Principles
96 Developments in Physical Adsorption
117 Adsorption Technology
134 Gas Purification by Adsorption
152 Adsorption and Ion Exchange
219 Recent Advances in Adsorption and Ion Exchange

AEROSPACE

33 Rocket and Missile Technology
40 Applications of Plastic Materials in Aerospace
52 Chemical Engineering Techniques in Aerospace
61 Aerospace Chemical Engineering
68 Aerospace Life Support

BIOENGINEERING

66 Chemical Engineering in Medicine
69 Bioengineering and Food Processing
84 The Artificial Kidney
86 Bioengineering . . . Food
93 Engineering of Unconventional Protein Production
99 Mass Transfer in Biological Systems
108 Food and Bioengineering—Fundamental and Industrial Aspects
114 Advances in Bioengineering
132 Engineering of Food Preservation and Biochemical Processes
158 Biochemical Engineering—Energy, Renewable Resources and New Foods
163 Water Removal Processes: Drying and Concentration of Foods and Other Materials
172 Food, pharmaceutical and bioengineering—1976/77
181 Biochemical Engineering Renewable Sources of Energy and Chemical Feedstocks

CRYSTALLIZATION

95 Crystallization from Solution and Melts
110 Factors Influencing Size Distribution
121 Nucleation Phenomena in Growing Crystal Systems
153 Analysis and Design of Crystallization Processes
193 Design Control and Analysis of Crystallization Processes

DRAG REDUCTION

111 Drag Reduction
130 Drag Reduction in Polymer Solutions

ENERGY

Conversion and Transfer

5 Heat Transfer, Atlantic City
9 Heat Transfer, Research Papers for 1954
17 Heat Transfer, St. Louis
18 Heat Transfer, Louisville
29 Heat Transfer, Chicago
30 Heat Transfer, Storrs
32 Heat Transfer, Buffalo
41 Heat Transfer, Houston
57 Heat Transfer, Boston
59 Heat Transfer, Cleveland
64 Heat Transfer, Los Angeles
75 Energy Conversion Systems
79 Heat Transfer with Phase Change
82 Heat Transfer—Seattle
87 Advances in Cryogenic Heat Transfer
92 Heat Transfer—Philadelphia
102 Heat Transfer—Minneapolis
113 Convective and Interfacial Heat Transfer
118 Heat Transfer—Tulsa
119 Commercial Power Generation
131 Heat Transfer: Fundamentals and Industrial Applications
138 Heat Transfer—Research and Design
162 Energy and Resource Recovery from Industrial and Municipal Solid Wastes
174 Heat Transfer: Research and Application
189 Heat Transfer—San Diego 1979
199 Heat Transfer—Orlando 1980
202 Transport with Chemical Reactions
208 Heat Transfer—Milwaukee 1981
216 Processing of Energy and Metallic Minerals

Nuclear Engineering

11 Part I (Ann Arbor)
12 Part II (Ann Arbor)
13 Part III (Ann Arbor)
19 Part IV
22 Part V
23 Part VI
27 Part VII
28 Part VIII
47 Part X
51 Part XI
53 Part XIII
56 Part XIV
60 Part XII
65 Part XV
68 Part XVI
71 Part XVII
80 Part XVIII
83 Part XIX
94 Part XX
104 Part XXI
106 Part XXII
119 Commercial Power Generation
123 Part XXIII
154 Radioactive Wastes from the Nuclear Fuel Cycle
164 Nuclear, Solar, and Process Heat Transfer—St. Louis 1976
168 Heat Transfer in Thermonuclear Power systems
169 Developments in Uranium Enrichment
191 Nuclear Engineering Questions Power Reprocessing, Waste, Decontamination Fusion
221 Recent Developments in Uranium Enrichment

ENVIRONMENT

35 Pollution and Environmental Health
45 Pollution Control Engineering
78 Water Reuse
90 Water—1968
97 Water—1969
107 Water—1970
115 Important Chemical Reactions in Air Pollution Control
122 Chemical Engineering Applications of Solid Waste Treatment
124 Water—1971
126 Air Pollution and its Control
129 Water—1972
133 Forest Products and the Environment
136 Water—1973
137 Recent Advances in Air Pollution Control
139 Advances in Processing and Utilization of Forest Products
144 Water—1974: I. Industrial Wastewater Treatment
145 Water—1974: II. Municipal Wastewater Treatment
146 Forest Product Residuals
147 Air: I. Pollution Control and Clean Energy
148 Air: II. Control of NO_x and SO_x Emissions
149 Trace Contaminants in the Environment
150 Advances in Interfacial Phenomena of Particulate/Solution/Gas Systems; Applications to Flotation Research
151 Water—1975
156 Air Pollution Control and Clean Energy
157 New Horizons for the Chemical Engineer in Pulp and Paper Technology
165 Dispersion and Control of Atmospheric Emissions, New-Energy-Source Pollution Potential
166 Water—1976: I. Physical, Chemical Wastewater Treatment
167 Water—1976: II. Biological Wastewater Treatment
170 Intermaterials Competition in the Management of Shrinking Resources
171 What the Filterman Needs to Know About Filtration
175 Control and Dispersion of Air Pollutants: Emphasis on NO_x and Particulate Emissions
177 Energy and Environmental Concerns in the Forest Products Industry
184 Advances in the Utilization and Processing of Forest Products
188 Control of Emissions from Stationary Combustion Sources Pollutant Detection and Behavior in the Atmosphere
190 Water—1978
195 The Role of Chemical Engineering in Utilizing the Nation's Forest Resources
196 Implications of the Clean Air Amendments of 1977 and of Energy Considerations for Air Pollution Control
197 Water—1979
198 Fundamentals and Applications of Solar Energy
200 New Process Alternatives in the Forest Products Industries
201 Emission Control from Stationary Power Sources: Technical, Economic and Environmental Assessments
207 The Use and Processing of Renewable Resources—Chemical Engineering Challenge of the Future
209 Water—1980
210 Fundamentals and Applications of Solar Energy II
211 Research Trends in Air Pollution Control: Scrubbing, Hot Gas Clean-up, Sampling and Analysis
213 Three Mile Island Cleanup

FLUIDIZATION

38 Fluidization
62 Fluid Particle Technology
67 Fluidized Bed Technology
101 Fundamental Processes in Fluidized Beds
105 Fluidization Fundamentals and Application
116 Fluidization: Fundamental Studies Solid-Fluid Reactions, and Applications
128 Fluidized Bed Fundamentals and Applications
141 Fluidization and Fluid-Particle Systems
161 Fluidization Theories and Applications
176 Fluidization Application to Coal Conversion Processes
205 Recent Advances in Fluidization and Fluid-Particle Systems
222 Fluidization and Fluid-Particle Systems: Theories And Applications

HISTORY OF CHEMICAL ENGINEERING

100 The History of Penicillin Production

ION EXCHANGE

14 Ion Exchange
24 Adsorption, Dialysis, and Ion Exchange
152 Adsorption and Ion Exchange
179 Adsorption and Ion Exchange Separations

(Continued on facing page)

ISBN 0-8169-0249-3